# Hazardous Waste Management

# Hazardous Waste Management

**Jaden Evans**

Larsen & Keller
www.larsen-keller.com

Hazardous Waste Management
Jaden Evans
ISBN: 978-1-64172-673-3 (Hardback)



Larsen & Keller
Published by Larsen and Keller Education,
5 Penn Plaza,
19th Floor,
New York, NY 10001, USA

**Cataloging-in-Publication Data**

Hazardous waste management / Jaden Evans.
p. cm.
Includes bibliographical references and index.
ISBN 978-1-64172-673-3
1. Hazardous wastes. 2. Hazardous wastes--Management. 3. Refuse and refuse disposal.
4. Pollution prevention. I. Evans, Jaden.
TD1030 .H39 2022
628.42--dc23



For more information regarding Larsen and Keller Education and its products, please visit the publisher's website www.larsen-keller.com

# Table of Contents


**Preface** VII

Chapter 1 **Hazardous Waste and its Types** 1
a. Characteristic Hazardous Waste 3
b. Listed Hazardous Waste 5
c. Universal Hazardous Waste 10
d. Mixed Hazardous Waste 12

Chapter 2 **Understanding Hazardous Waste Management** 41
a. Radioactive Waste Management 51

Chapter 3 **Hazardous Waste Treatment Technologies** 61
a. Physical Treatment of Hazardous Waste 61
b. Chemical Treatment of Hazardous Waste 67
c. Physicochemical Treatment 96
d. Biological Treatment of Hazardous Waste 98
e. Stabilization/Solidification of Hazardous Waste 104

Chapter 4 **Disposal and Storage of Hazardous Waste** 109
a. Methods to Dispose Hazardous Waste 118
b. Treatment Methods used in Disposal 122
c. Hazardous Waste Recycling 126
d. Hazardous Waste Storage 194

Chapter 5 **Impacts of Hazardous Waste** 197
a. Impacts of Hazardous Waste on Human Health 200
b. Environmental Impact of Household Hazardous Waste 202

**Permissions**

**Index**

# Preface

Hazardous waste refers to the waste which can cause significant threat to the environment or public health. These cannot be disposed off by the use of regular means. Generally, hazardous wastes possess a few characteristic traits. A few of these are reactivity, toxicity, corrosivity and ignitability. Various types of hazardous waste include pesticides and other garden chemicals, radioactive waste, pharmaceutical waste, etc. The management of hazardous waste involves stabilizing it before disposing it. There are various ways in which hazardous waste can be managed such as by recycling them into new products, solidifying and stabilizing them by turning them into portland cement and pyrolysis. They can also be destroyed using the process of incineration. They can also serve as fuel when incinerated. The topics included in this book on hazardous waste management are of utmost significance and bound to provide incredible insights to readers. It presents researches and studies performed by experts across the globe. This book is an essential guide for both academicians and those who wish to pursue this discipline further.

To facilitate a deeper understanding of the contents of this book a short introduction of every chapter is written below:

Chapter 1- Any waste material that poses probable and potential threats to the environment is termed as hazardous waste. It includes plastics, metals, paint and solvents, insecticides, herbicides, fungicides, aerosols, etc. This is an introductory chapter which will briefly introduce all the significant aspects of hazardous waste and its types.

Chapter 2- A system which is used for the management, collection, treatment and disposal of hazardous waste to protect the environment is called a hazardous waste management system. Radioactive waste management refers to treating, storing, and disposing liquid, airborne, and solid effluents from the nuclear industry's operations. This chapter has been carefully written to provide an easy understanding of the varied facts of hazardous waste management.

Chapter 3- There are various methods and technologies which are used to treat hazardous waste. Pyrolysis, gastification, incineration, bioremediation and landfill are some of the different chemical, biological and physical treatment methods. This chapter discusses these techniques and methods related to hazardous waste treatment in detail.

Chapter 4- The process of discarding hazardous waste after its treatment is termed as hazardous waste disposal. Hazardous waste recycling includes glass recycling, plastic recycling, paper recycling, metal recycling, etc. This chapter delves into the methods and related concepts of hazardous waste disposal and storage to provide an in-depth understanding of the subject.

Chapter 5- Hazardous waste has many adverse impacts on the human health and environment. Air, soil, water, wildlife, etc. are some of the biotic and abiotic factors of the environment which are severely affected by the amounts of hazardous waste. The topics elaborated in this chapter will help in gaining a better perspective about the impacts of hazardous waste.

I would like to share the credit of this book with my editorial team who worked tirelessly on this book. I owe the completion of this book to the never-ending support of my family, who supported me throughout the project.

**Jaden Evans**

1

# Hazardous Waste and its Types

Any waste material that poses probable and potential threats to the environment is termed as hazardous waste. It includes plastics, metals, paint and solvents, insecticides, herbicides, fungicides, aerosols, etc. This is an introductory chapter which will briefly introduce all the significant aspects of hazardous waste and its types.

Hazardous waste is waste that is dangerous or potentially harmful to our health or the environment. Hazardous wastes can be liquids, solids, gases, or sludges. They can be discarded commercial products, like cleaning fluids or pesticides, or the by-products of manufacturing processes.

The universe of hazardous wastes is large and diverse. Hazardous wastes can be liquids, solids, or contained gases. They can be the by-products of manufacturing processes, discarded used materials, or discarded unused commercial products, such as cleaning fluids (solvents) or pesticides. In regulatory terms, a hazardous waste is a waste that appears on one of the four RCRA1 hazardous wastes lists (the F-list, K-list, P-list, or U-list) or that exhibits one of the four characteristics of a hazardous waste - ignitability, corrosivity, reactivity, or toxicity. However, materials can be hazardous wastes even if they are not specifically listed or don't exhibit any characteristic of a hazardous waste. For example, "used oil," products which contain materials on California's M-list, materials regulated pursuant to the mixture or derived-from rules, and contaminated soil generated from a "clean up" can also be hazardous wastes.

## Hazardous Characteristics

## Industrial Wastes

Waste generated from industrial sources can have non-hazardous and hazardous components, with non-hazardous waste usually representing the greater part of the volume. The hazardous component of this waste is relatively small in volume.

This type of waste was identified as hazardous waste when proceeds toxicity test, corrosively test, ignitability test, and some special character test. As a hazardous pollutant, it may impose serious impacts on surrounding environment and such impacts should be quantitatively examined to assess the influence on human health.

## Household Waste

Households generate small quantities of hazardous wastes such as oil-based paints, paint thinners, wood preservatives, pesticides, insecticides, household cleaners, used motor oil, antifreeze, and batteries. It has been estimated that household hazardous waste in industrialized countries such as the United States accounts for a total of about 0.5% (by weight) of all waste generated at home, while in most developing countries, the percentage probably is even lower.

## Biomedical Waste

There are some of hazardous medical and dental wastes that, when disposed improperly, could cause harm to the environment. It also presents an occupational health hazards to the healthcare personnel who handle these wastes at the point of generation and those involved with their management, that is, segregation, storage, transport, treatment, and disposal.

Healthcare waste that is capable of producing injury or disease including many sorts of hazardous wastes such as:

- Infectious waste: Which contain pathogens namely bacteria, viruses, fungi, or parasites in concentrations sufficient to cause disease in susceptible hosts. Cultures and stocks of infectious agents from laboratory work; tissues and dressing generated from autopsies, surgeries, and treatment of infected patients and animals; materials or equipment in contact with blood and infected body fluids.
- Pathological waste: Including tissue, organs, body parts, human fetuses, and animal carcasses, blood and body fluids.
- Sharps: It comprise syringes, needles, scalpels, saws, infusion sets, knives, blades, broken glass, or other items that can cause cut or puncture wounds.
- Pharmaceutical waste: It covers expired, unused, spilt, and contaminated pharmaceutical products, drugs, vaccines, and sera that are no longer required and need to be disposed of in appropriate manner.
- Genotoxic waste: This type combines cytostatic drugs, vomit, urine, or feces from the patients treated with cytotoxic drugs, chemicals, and radioactive materials. Genotoxic waste has mutagenic, teratogenic, and carcinogenic properties.
- Chemical waste: Discarded solid, liquid, or gaseous chemicals should be considered as hazardous if it is toxic, corrosive, inflammable, or reactive.
- Waste with high content of heavy metals: Mercury (thermometers, blood pressure gauges, amalgam), cadmium (discarded batteries), and lead (reinforced wood panels for radiation proofing in radiology department) generated from hospitals could be represented as a subcategory of hazardous chemical waste.
- Radioactive waste: The use of radioisotopes in vitro analysis of body tissues and fluids, in vivo organ imaging, tumor localization, and treatment and various clinical studies involving certain radionuclides need to be specially managed in a centralized treatment facility for radioactive wastes.

## Radioactive Waste

Nuclear applications have been rapidly developed recently, and several nuclear power plants started to work throughout the world. The potential impact of released radioactive contaminants into the environment has received growing attention due to nuclear accidents. Contamination of soil and water by radionuclides due to natural processes, global fallout from nuclear weapon testing, discharges from nuclear installations, disposal of nuclear waste, and occasional nuclear accidents poses serious problems to biological systems. Radioactive waste includes a variety of radionuclides and occurs in a variety of physical and chemical forms. It can be generally classified as low-/intermediate-level radioactive waste and high-level radioactive waste.

Radioactive waste, arising from civilian nuclear activities as well as from weapon activities, poses a potential problem for handling and saving the environment for coming generations.

Radioactive waste includes a variety of radionuclides and occurs in a variety of physical and chemical forms. It can be generally classified as low-/intermediate-level radioactive waste and high-level radioactive waste. Nuclear research establishments include, for example, waste containing different organic components, toxic or chemically aggressive constituents, radionuclides with specific properties (high mobility, high chemical activity, volatile elements, etc.), waste difficult for treatment and not appropriate for direct immobilization (e.g., spent organic ion exchange resins and spent liquid scintillation cocktails). For such waste, application of conventional treatment and conditioning options may not be efficient and appropriate in terms of economy, safety, and performance characteristics. In many cases, such wastes are stored awaiting an appropriate treatment and conditioning solution.

The primary sources of radioactive wastes in a country without nuclear fuel cycle activities are nuclear research, production of radioisotopes, application of radioisotopes, and decontamination and decommission of nuclear installations.

## Characteristic Hazardous Waste

Characteristic wastes are wastes that exhibit measurable properties which indicate that a waste poses enough of a threat to deserve regulation as hazardous waste. EPA tried to identify characteristics which, when present in a waste, can cause death or illness in humans or lead to ecological damage. The characteristics are an essential supplement to the hazardous waste listings. For example, some wastes may not meet any listing description because they do not originate from specific industrial or process sources, but the waste may still pose threats to human health and the environment. As a result, a facility is also required to determine whether such a waste possesses a hazardous property (i.e., exhibits a hazardous waste characteristic). The characteristics are applied to any waste from any industry.

Even if a waste does meet a hazardous waste listing description, the facility must still determine if the waste exhibits a characteristic. If such listed wastes do exhibit a characteristic, the waste poses an additional hazard to human health and the environment, and may necessitate additional regulatory precautions. For example, wastes that are both listed and characteristic may have more extensive LDR requirements, than those that are only listed.

EPA decided that the characteristics of hazardous waste should be detectable by using a standardized test method or by applying general knowledge of the waste's properties. Given these criteria, EPA established four hazardous waste characteristics: Ignitability, Corrosivity, Reactivity, Toxicity.

## Ignitability

The ignitability characteristic identifies wastes that can readily catch fire and sustain combustion. Many paints, cleaners, and other industrial wastes pose such a hazard. Liquid and nonliquid wastes are treated differently by the ignitability characteristic.

Most ignitable wastes are liquid in physical form. EPA selected a flash point test as the method for determining whether a liquid waste is combustible enough to deserve regulation as hazardous. The flash point test determines the lowest temperature at which the fumes above a waste will ignite when exposed to flame.

The ignitability characteristic identifies wastes that can readily catch fire and sustain combustion.

Many wastes in solid or nonliquid physical form (e.g., wood, paper) can also readily catch fire and sustain combustion, but EPA did not intend to regulate most of these nonliquid materials as ignitable wastes. A nonliquid waste is considered ignitable if it can spontaneously catch fire or catch fire through friction or absorption of moisture under normal handling conditions and can burn so vigorously that it creates a hazard. Certain compressed gases are also classified as ignitable. Finally, substances meeting the Department of Transportation's definition of oxidizer are classified as ignitable wastes. Ignitable wastes carry the waste code D001 and are among the most common hazardous wastes. The regulations describing the characteristic of ignitability are codified at 40 CFR 261.21.

## Corrosivity

The corrosivity characteristic identifies wastes that are acidic or alkaline (basic) and can readily corrode or dissolve flesh, metal, or other materials.

The corrosivity characteristic identifies wastes that are acidic or alkaline (basic). Such wastes can readily corrode or dissolve flesh, metal, or other materials. They are also among the most common hazardous wastes. An example is waste sulfuric acid from automotive batteries. EPA uses two criteria to identify liquid and aqueous corrosive hazardous wastes. The first is a pH test. Aqueous wastes with a pH greater than or equal to 12.5 or less than or equal to 2 are corrosive. A liquid waste may also be corrosive if it has the ability to corrode steel under specific conditions. Physically solid, nonaqueous wastes are not evaluated for corrosivity. Corrosive wastes carry the waste code D002. The regulations describing the corrosivity characteristic are found at 40 CFR 261.22.

## Reactivity

The reactivity characteristic identifies wastes that readily explode or undergo violent reactions.

The reactivity characteristic identifies wastes that readily explode or undergo violent reactions. Common examples are discarded munitions or explosives. In many cases, there is no reliable test method to evaluate a waste's potential to explode or react violently under common handling conditions. Therefore, EPA uses narrative criteria to define most reactive wastes and requires waste handlers to use their best judgment in determining if a waste is sufficiently reactive to be regulated. This is possible because reactive hazardous wastes are relatively uncommon and the dangers that they pose are believed to be well known to the few waste handlers who deal with them.

A waste is reactive if it meets any of the following criteria:

- It can explode or violently react when exposed to water or under normal handling conditions.
- It can create toxic fumes or gases when exposed to water or under normal handling conditions.
- It meets the criteria for classification as an explosive under DOT rules.
- It generates toxic levels of sulfide or cyanide gas when exposed to a pH range of 2 through 12.5. Wastes exhibiting the characteristic of reactivity are assigned the waste code D003.

The reactivity characteristic is described in the regulations at 40 CFR 261.23.

### Toxicity

When hazardous waste is disposed of in a land disposal unit, toxic compounds or elements can leach into underground drinking water supplies and expose users of the water to hazardous chemicals and constituents. EPA developed the toxicity characteristic (TC) to identify wastes likely to leach dangerous concentrations of toxic chemicals into ground water.

In order to predict whether any particular waste is likely to leach chemicals into ground water at dangerous levels, EPA designed a lab procedure to replicate the leaching process and other conditions that occur when wastes are buried in a typical municipal landfill. This lab procedure is known as the Toxicity Characteristic Leaching Procedure (TCLP).

The regulations require a facility to apply the TCLP to its hazardous waste samples in order to create a liquid leachate. This leachate would be similar to the leachate generated by a landfill containing a mixture of household and industrial wastes. Once this leachate is created via the TCLP, the waste handler must determine whether it contains any of 40 different toxic chemicals in amounts above the specified regulatory levels. These regulatory levels are based on ground water modeling studies and toxicity data that calculate the limit above which these common toxic compounds and elements will threaten human health and the environment. If the leachate sample contains a concentration above the regulatory limit for one of the specified chemicals, the waste exhibits the toxicity characteristic and carries the waste code associated with that compound or element. The regulations describing the toxicity characteristic cannot be individually measured, the regulatory level for total cresols is used.

## Listed Hazardous Waste

EPA has determined that some particular wastes are hazardous. These hazardous wastes are included in EPA's "F", "K", "P", or "U" lists. Each list has a different focus and can be found in 40 CFR § 261.31 through 261.33.

## F-listed Wastes

F-listed wastes are hazardous wastes from non-specific sources. Most of the F-list is fairly easy to understand. Some examples follow.

F019 is the waste code for wastewater treatment sludges from the chemical conversion coating of aluminum. If a generator uses a chemical conversion coating process for aluminum, and if the wastewater is treated on-site, then the sludge from that wastewater treatment is a listed hazardous waste with the F019 waste code.

F006 is the waste code for wastewater treatment sludges from electroplating operations except from the following processes: (1) Sulfuric acid anodizing of aluminum; (2) tin plating on carbon steel; (3) zinc plating (segregated basis) on carbon steel; (4) aluminum or zinc-aluminum plating on carbon steel; (5) cleaning/stripping associated with tin, zinc and aluminum plating on carbon steel; and (6) chemical etching and milling of aluminum. Therefore, if a generator has a chrome-plating operation and the wastewater is treated on-site, the sludge from the treatment of that wastewater is a listed hazardous waste with the waste code F006.

The most common F-listed waste codes used in Kansas are F001, F002, F003, F004, and F005. These are the waste codes for specific spent (i.e., used) solvents, and they appear frequently because many wastes generated from painting operations and parts washers contain solvents. Each of these spent solvent listings is defined by how a specific solvent is used and, with the exception of the F003 listing, the percentage of a specific constituent in the solvent before it is used. Please be aware that more than one of these F-listings may apply to a single waste stream.

## F001, F002, F004 and F005 Waste Codes

When a solvent mixture contains a total of 10% or more of one spent solvent, or a combination of spent solvents, from the F001, F002, F004, and F005 listings, then all appropriate F-listings will apply to the mixture. In order to clarify how to apply these listings, the following two scenarios are presented.

Scenario 1: A generator has a parts washer with a solvent that has become too dirty to use. The generator must now make a hazardous waste determination on the spent solvent so that it can be properly disposed of. The generator looks at the SDS for the solvent and it lists: trichloroethylene (65%) and methylene chloride (35%). These solvents are found both on the F001 and F002 lists. The parts washer is not part of a degreasing (large-scale cleaning) operation, so the F001 listing does not apply. The F002 listing does apply to this waste, since the combination of trichloroethylene and methylene chloride makes up 10% or more of the mixture (in this case 100%). Therefore, this mixture is a listed hazardous waste carrying the F002 waste code.

Scenario 2: A generator uses methyl ethyl ketone (MEK) to thin paint. The waste paint is poured into a 55-gallon drum. The generator also cleans the paint gun and lines with MEK, but this waste is put in a 15- gallon drum that is kept segregated from the waste paint. The generator looks at the F005 listing and sees that MEK that has been used as a solvent is a listed hazardous waste carrying the F005 waste code. Therefore the 15-gallon drum of spent solvent (MEK) is a listed hazardous waste (F005). The 55-gallon drum of waste paint is not a listed hazardous waste because the MEK was used as an ingredient in the paint; it was not used as a solvent. However, the waste paint may be a hazardous waste if it exhibits the characteristic of ignitability (D001) and/or MEK toxicity (D035). An analysis of a representative sample would be necessary to determine if the waste paint is characteristic for MEK toxicity.

## F003 Waste Code

The definition of an F003-listed waste is more complicated. There are two ways that a spent solvent mixture can meet the F003 listing:

- The mixture contains only F003 constituents that are pure or technical grade; or
- The mixture contains one or more F003 constituents and a total of 10% or more of the other listed solvents (F001, F002, F004, and/or F005) before use.

In either of the above cases, the spent solvent mixture must be ignitable (have a flashpoint of less than 140 °F) at the point of generation in order for the F003 listing to apply. If the spent solvent mixture is not ignitable at the point of generation, then the F003-listing will not apply.

Below are five scenarios that demonstrate how to apply the F003 listing.

Scenario 1: A generator uses a solvent for cleaning. The SDS shows acetone (99.9%) and water (0.1%). This would be considered a technical grade for acetone.

- The generator looks at the F003 list and sees that acetone is on the list.
- The generator determines that the spent solvent is ignitable because it has a flash point of less than 140 °F at the point of generation.

Therefore, the spent acetone will be a listed hazardous waste carrying the waste code F003 because the spent solvent is composed only of a pure or technical grade (essentially 100%) F003-listed solvent and is ignitable at the point of generation.

In most painting and parts washer applications, F003-listed solvents are not used in pure or technical grade form. These solvents are usually mixed/blended with other solvents. The next three examples will show how to apply the F003 listing for a solvent mixture.

Scenario 2: A generator uses a solvent for cleaning. The SDS for the solvent shows acetone (89%), toluene (6%), and MEK (5%).

Determining the applicable waste codes can be broken down into the following steps:

- The generator looks at the F005 list and sees that toluene and MEK are both on the list. The combined percentage of the toluene and MEK is 11%, which meets the listing definition of a total of 10% or more. Therefore, the spent solvent is a listed hazardous waste with the waste code F005.
- The generator also looks at the F003 list and sees that acetone is on the list. The generator sees that, by definition, the listing applies if the spent solvent contains one or more of the solvents on the F003 list (in this case, acetone) and a total of 10% or more of one or more of the solvents listed on the F001, F002, F004, and F005 lists (in this case, 11% of F005-listed

solvents). Therefore, if the spent solvent is ignitable at the point of generation it will meet the definition of a F003-listed waste.

- The generator determines that the spent solvent is ignitable because it has a flash point of less than 140 °F at the point of generation. Therefore, the spent solvent meets the definition of a F003-listed waste.

The spent solvent in this scenario would carry both the F003 and F005 waste codes. The important thing to remember is that the percentage of the F003-listed solvent does not matter as long as there is a total of 10% or more of the other F-listed solvents (F001, F002, F004, and/or F005) and the spent solvent is ignitable at the point of generation.

In the above example, if the generator had determined that the spent solvent was not ignitable at the point of generation, then the F003 listing would not have applied and the only applicable waste code would have been F005.

Scenario 3 – The generator from Scenario 2 recycles his solvent in a distillation unit on-site. Still bottoms (i.e., pancakes, still cakes, pucks, solids) are generated from this process. These solids will also be a listed hazardous waste carrying the F003 and F005 waste codes because the definition for both F003 and F005 listings states that they include "still bottoms from the recovery of these spent solvents and spent solvent mixtures." This is the only case where it doesn't matter whether the waste is ignitable at the point of generation; the F003 listing applies entirely because of the still bottoms definition (40 CFR § 261.31).

Scenario 4 – A generator uses a solvent for cleaning. The SDS for the solvent shows acetone (65%) and water (35%). The generator looks at the F003 list and sees that acetone is on it. However, based on the SDS, the solvent is not pure or technical grade acetone (because it does not contain essentially 100% F003-listed solvent). Also, the solvent does not contain any other F-listed constituents (F001, F002, F004, or F005). Therefore, the spent solvent does not meet the definition of an F-listed spent solvent, and does not carry any of those listed waste codes. However, if the spent solvent is ignitable at the point of generation it will carry the characteristic waste code for ignitability, D001.

Scenario 5 – A generator uses a solvent for cleaning paint brushes. The SDS for the solvent shows toluene (65%), methylene chloride (5%), xylene (5%), and water (25%).

Determining the applicable waste codes can be broken down into the following steps:

- The generator finds toluene on the F005 list.

- The generator finds methylene chloride on both the F001 and the F002 lists. The F001 listing applies to the listed solvents used in large-scale industrial degreasing operations. The F002 listing applies to the listed solvents used for other applications than large-scale industrial degreasing. In this example, F002 is the applicable waste code because the generator is using the solvent to clean paint brushes, which is not large-scale, industrial degreasing.

- The generator combines the percentages of toluene (65%) and methylene chloride (5%), for a total percentage of 70%. By both the F002 and F005 definitions, the spent solvent would be listed for both waste codes because the solvent contains 10% or more of a solvent, or a combination of solvents, listed in F001, F002, F004, and F005.

- The generator finds xylene in the F003 list. The generator sees that, by definition, the listing applies if the waste (spent) solvent contains one or more of the solvents on the F003 list (in this case, xylene) and a total of 10% or more of one or more of the solvents listed in the F001, F002, F004, and F005 lists (in this case, a total of 70% of F002- and F005-listed solvents). Therefore, if the spent solvent is ignitable at the point of generation it will meet the definition of a F003-listed waste.
- The generator determines that the spent solvent is ignitable because it has a flash point of less than 140 °F at the point of generation. Therefore, the spent solvent meets the definition of an F003- listed waste.

The spent solvent in this scenario will carry the F002, F003, and F005 waste codes.

In the above example, if the generator had determined that the spent solvent was not ignitable at the point of generation, then the F003 listing would not have applied and the only applicable waste codes would have been F002 and F005.

It is important to note that on the one-time LDR notification for a spent solvent that carries the F003 waste code, the D001 waste code must be included in addition to the F003 waste code. This ensures that the spent solvent will be treated for the characteristic of ignitability.

## K-listed Wastes

K-listed wastes are hazardous wastes from specific sources. A generator must utilize the specific process listed in the definition in order to have a K-listed waste. An example of K-listed hazardous waste is distillation bottoms from the production of acetaldehyde from ethylene, which carries a waste code of K009.

## P- and U-listed Wastes

P- and U-listed wastes are discarded commercial chemical products, off-specification species, container residues, and spill clean-up materials from any of these that must be disposed for some reason. The difference between the "P" and "U" lists lies in the toxicity of the product. The acutely toxic products are on the P-list and less toxic products are on the U-list. The product will only become a waste at the point the generator no longer has a use for it and determines that it must be discarded (or if it is spilled and cleaned up).

P- and U-listed hazardous wastes are unused (not spent) materials. If the discarded material is a solution, the generator should look at the active ingredient of the solution. Some common examples of P- and Ulisted wastes are:

- Discarded nicotine patches (P075)
- Warfarin (P001)
- Acetone (U002)

- Benzene (U019)
- Toluene (U220)

Remember that a material must be unused to meet the definition of P- or U-listed waste.

P-listed wastes are less common, but due to their acute toxicity they are regulated at a much smaller quantity than other hazardous wastes. A facility that generates 2.2 pounds or more in a calendar month, or accumulates at any time 2.2 pounds or more, of P-listed hazardous waste, is regulated as a Large Quantity Generator.

# Universal Hazardous Waste

Universal Hazardous Waste comes primarily from consumer products containing mercury, lead, cadmium and other substances that are hazardous to human health and the environment. These items cannot be discarded in household trash nor disposed of in landfills. Examples of universal waste are batteries, fluorescent tubes, and many electronic devices.

## Types of Universal Hazardous Waste

### Universal Waste – Batteries

This includes discarded primary (non-rechargeable) and secondary (rechargeable) batteries that contain elements such as cadmium, lead, or mercury, which would render them RCRA-hazardous.

Examples are nickelcadmium (Ni-Cad), sealed lead-acid, or mercury-oxide batteries. Lead-acid batteries (such as automotive batteries) that are generated, transported, or collected to be reclaimed, or regenerated, but not reclaimed where stored, under provisions of 40 CFR 266, Subpart G, “Spent Lead-Acid Batteries Being Reclaimed,” do not need to be managed as universal waste. However, waste lead-acid batteries not managed, or eligible for management, under 40 CFR 266, Subpart G, are subject to the Universal Waste Rule requirements. Lead-acid batteries that are stored at facilities that reclaim them are subject to RCRA regulation as specified in 40 CFR 266.80(b).

Many commonly generated waste batteries, such as dry cell zinc-carbon and alkaline (“long life”) batteries, typically do not contain appreciable amounts of the hazardous elements of concern, and hence would not be required to be managed as universal waste. However, they may be managed along with universal waste batteries, and this is encouraged in the interest of diverting them from less desirable disposal destinies such as incineration or disposal in solid waste landfills.

### Universal Waste – Pesticides

Several classes of discarded pesticides that would otherwise be regulated as characteristic or listed hazardous waste may be eligible for management under the Universal Waste Rule:

- Stocks of unused suspended or canceled pesticides that are subject to a voluntary or mandatory recall under the section 19(b) of the Federal Insecticide, Fungicide, and Rodenticide

Act (FIFRA), or a voluntary recall by a registrant of a pesticide that is not in compliance with FIFRA;

- Stocks of other unused pesticide products that are collected and managed as part of a waste pesticide collection program.

Pesticides not subject to the prescribed recalls may have to be managed as hazardous waste when discarded. Farmers managing and disposing such pesticides in accordance with the hazardous waste exclusion in 40 CFR 262.70, and complying with prescribed disposal instructions, are not subject to the Universal Waste Rule. Recalled or unused pesticides are not considered waste until a decision to discard them has been made. However, they remain subject to the requirements of FIFRA.

## Universal Waste – Mercury Containing Equipment

This category was added to the Federal Rule at 40 CFR 273 on August 5, 2005. It includes devices, items, or articles which contain elemental mercury that is integral to their functions and which would otherwise be regulated as a hazardous waste when discarded by virtue of exhibiting the toxicity characteristic (TC) for mercury (hazardous waste code D009).

Some examples of such items include mercury-containing thermostats (formerly a separate universal waste category in the original rule), thermometers, barometers, mercury switches, and certain types of meters, regulators, and gauges, in which elemental mercury is contained in ampules or otherwise enclosed and can be managed intact. This category does not include cathode ray tubes or other types of electronic equipment in which elemental mercury is not integral to function, nor does it include mercury waste that is generated as a byproduct through the process of manufacturing or treatment.

## Universal Waste – Lamps

This category was added to the Federal Rule at 40 CFR 273 on July 6, 1999. The universal waste lamp category is slightly broader than its predecessor in that it includes lamps that are hazardous for any characteristic, not just for mercury. However, fluorescent light bulbs remain the most common item in this category of universal waste. This category does not include associated light fixture components such as ballasts.

Hazardous waste lamps become subject to this rule if they are hazardous waste under 40 CFR 261, and when they are permanently removed from a fixture or determined to be discarded.

A note about crushing: In adopting the Federal category of universal waste lamps, the Indiana rule has added a prohibition against intentionally breaking or crushing waste lamps that are managed under the reduced requirements of the Universal Waste Rule.

This is a consequence of the interpretation that crushing of hazardous waste bulbs is treatment, which is explicitly prohibited under the Universal Waste Rule (40 CFR 273.11(b) and 273.31(b)). The only circumstance where hazardous waste lamps may be crushed is when they are managed as fully regulated hazardous waste (rather than as universal waste) which is treated (i.e., crushed) in tanks or containers by the generator of the waste under the implied generator treatment allowance in 40 CFR 262.34, and in conformance with all applicable hazardous waste management standards. Check with local government to find regulations like this applicable in your state.

## Mixed Hazardous Waste

Mixed hazardous waste is waste which falls into two or more different categories of hazardous materials. Examples include radioactive contaminated phenol/chloroform, or blood labeled with a radionuclide. Special disposal procedures must be followed with mixed wastes to prevent handlers from receiving an exposure from one hazard while treating another. Also, disposal regulations require different disposal methods for mixed wastes.

### Biodegradable and Non-biodegradable Waste

Technology has improved the quality of life and given birth to an advanced form of life. This has varying impact on the environment and other lives on the earth. For instance, water bottles made of plastic, which is convenient to use and carry but its disposal is a challenge for nature.

There are lots of waste materials which are degradable and non-degradable. Therefore, categorization and proper disposal of wastes are necessary. Waste materials can be classified as biodegradable and non-biodegradable. Risk and threats related to waste disposals can be easily avoided by the knowledge on types of wastes.

### Biodegradable

A biodegradable material can be defined as a material which can be decomposed by b bacteria or other natural organisms and not be adding to pollution.

Biodegradable wastes are such waste materials which are and can be degraded by natural factors like microbes (e.g. bacteria, fungi and few more), abiotic elements like temperature, UV, oxygen, etc. Some examples of such wastes are food materials, kitchen wastes, and other natural wastes. Microorganisms and other abiotic factors together break down complex substances into simpler organic matters which eventually suspend and fade into the soil. The whole process is natural which can be rapid or slow. Therefore the environmental issues and risks caused by biodegradable wastes are low.

Examples:

But the huge dumping of waste can raise some threats to life sooner or later. To avoid this, some people practice composting. In composting, the biodegradable wastes are dumped into a pit and covered for a period. Due to the action of microbes, they will decompose and will be used as manure for cultivation purpose. This will reduce the amount of waste at landfills.

### Biodegradable Waste

Biodegradable waste is a type of waste, typically originating from plant or animal sources, which may be degraded by other living organisms.

Biodegradable waste can be commonly found in municipal solid waste as green waste, food waste, paper waste, and biodegradable plastics. Other biodegradable wastes include human waste, manure, sewage, slaughterhouse waste.

## Non-biodegradable

A Non-Biodegradable material can be defined as a kind of substance which cannot be broken down by natural organisms and acts as a source of pollution.

Unlike biodegradable wastes, non-biodegradable cannot be easily handled. Non-biodegradable wastes are those who cannot be decomposed or dissolved by natural agents. They remain on earth for thousands of years without any degradation. Hence the threat caused by them is also more critical. A notable example is the plastics which are a commonly used material in almost every field. To give these plastics a long lasting effect, improved quality plastics are being put to use. This made them more temperature resistant and more durable even after use. Other examples are cans, metals, and chemicals for agricultural and industrial purposes. They are the main causes of air, water and soil pollution and diseases like cancer.

Since non-biodegradable wastes are not Eco-friendly, they need to be replaced. As a part of a development of alternatives, scientists have brought forward many ides like biodegradable plastics, etc. They incorporated some biodegradable materials with plastics and made them easily and rapidly degradable. But this is quite an expensive procedure.

## Non-biodegradable Waste

Waste which cannot be decomposed by biological process is known as "Non-biodegradable wastes". Most of the inorganic waste is non-biodegradable. Non-biodegradable wastes which can be recycled are known as "Recyclable waste" and those which cannot be recycled are known as "Non-recyclable waste".

Difference between Biodegradable and non-biodegradable:

| S. No | Biodegradable | Non Biodegradable |
|---|---|---|
| 1 | Degradation process in Biodegradable waste is is rapid. | Degradation process in Non-Biodegradable waste is slow. |
| 2 | Biodegradable waste is decomposed and degraded by microbes. | Non-Biodegradable waste is cannot be decomposed by microbes. |

| | | |
|---|---|---|
| 3 | Biodegradable waste are not accumulated but are used up in short time. | Non-Biodegradable waste often accumulate. |
| 4 | Biodegradable waste become part of bio-geochemical cycles and give back rapid turnover. | Most of Non-Biodegradable waste never enter into bio-geochemical cycles, very slow and toxic. |
| 5 | Biodegradable waste are used to produce energy manure, compost and biogas. | Non-Biodegradable waste can be separated and recycled but the process is very expensive. |

## Effects of Waste Material on Environment

Waste quantities are generally growing in all countries all around the world. Every year billions of tons of waste are generated. These wastes are result of activities in our homes, businesses and industries and disposal of all this large amount is an enormous environmental problem with many dimensions. Municipal, industrial and agricultural solid waste and biomass deposits cause large scale pollution of land and water. The generation of waste causes a loss of materials and energy and increase environmental costs on society for its collection, treatment and disposal. The impacts of landfill and incineration are significant because of their potential for greenhouse gas emissions (methane, carbon dioxide) and trans-boundary migration of organic micro-pollutants (dioxins and furans) and volatile heavy metals. Problems with waste are as old as human race.

Very soon humans realized that waste are a potential source of diseases and infections, so they dump their waste, which were totally biological, away from their settlements. The first organized municipal dump is in 500 BC outside ancient Athens in Greece, where regulations required waste to be dumped at least a mile from the city limits and covered with soil. Until industrialization of society waste was mostly organic, so they can decompose naturally. Later, mostly because of industrialization, urbanization, and developing of consumer society amount of waste increase very fast.

## Plastic

Plastics are chemically inert, that's a common belief.

The ability to be shaped into almost any object with practical value is what makes plastics so popular. Chairs, drums, food containers, bulletproof vests, packaging materials, bags, vehicles—almost anything around us can be constructed from plastics. Being lightweight, low in cost, and reasonably durable, plastics have replaced metals and glass in many areas of application.

Despite these desirable properties, plastics are major contributors to pollution due to their nature to degrade very slowly. In fact, their degradation takes years, decades or even centuries, making them practically non-degradable materials.

As plastic wastes continue to litter our lands and bodies of water, they continue to disrupt the ecosystem and lives of organisms in these environments. That's why countries have been pushing the principle of 3R—Reduce, Reuse and Recycle—to minimize the impact of plastics. Still, this has not deterred the increasing amount of plastic wastes we are generating every day. The Environmental Protection Agency or EPA noted that only around 12% of plastic materials gets recycled.

The researchers who made the call for reclassification focused their targets on four plastics believed to be the most harmful and hard to recycle: PVC (polyvinyl chloride); polystyrene, commonly known as Styrofoam; polyurethane; and polycarbonate. You can find these materials in wide applications, from pipes to bottles to bed mattresses.

## The Problem with Plastics

For a waste to be deemed as a "hazardous waste," it must pose dangers to human health or the environment. Due to these hazards, such wastes must be managed carefully and responsibly as improper disposal could lead to major health and environmental issues. In many countries, regulations are in place, and government agencies like EPA exist to ensure hazardous wastes are being handled properly.

Right now, plastics are classified as an ordinary solid waste, like food scraps and grass clippings. Under this classification, plastics are allowed to be disposed of in the garbage bin and ultimately into our landfills. Here are some arguments why plastics must be reclassified as hazardous wastes:

- Plastics are practically non-degradable. It could take up to 500 years to degrade certain types of plastics. As they remain intact for long periods of time, dumped plastics will continue to accumulate on the ground, which reduces the value and cultivability of lands. Plastic debris also tends to continue absorbing pollutants like toxic metals and residual pesticides.
- Plastics also clog sewage and pipelines. They block waterways or the seepage of water into the soil. Five trillion pieces of plastics (roughly 250,000 metric tons) find their way into the ocean, creating massive "plastic soups" as seen in the Atlantic and the Pacific (the latter known as the "Pacific Garbage Patch"). Apart from the ghastly sight of floating plastic debris, these wastes also cause harms to the aquatic environment and the living inhabitants there. As they float in water or settle into lands, plastics can be ingested by animals or other organisms, disrupting the food web balance.
- Direct burning as a way of disposing plastics is harmful as well. This method emit gases and fumes which may be harmful to human health. Not only that, burning plastics also emits carbon dioxide, a greenhouse gas that plays a role in global warming.
- Exposure to materials and processes involving plastics production pose harm to humans and other organisms. The process of manufacturing plastic products produce pollutants and other toxic by-products that can find their paths into waterways, ground or any part of the environment. Some raw materials used to make them are also known hazardous materials, like the monomers. For instance, workers are exposed to toxic vinyl chloride fumes (which is a known carcinogen) during PVC production.
- As a final product, plastics still deliver varying degrees of hazards. Certain additives (like Azo dyes) and residual ingredients (like residual monomers and Bisphenol A) are a threat to consumer health. These components tend to seep out during usage or leach into the ground when plastic wastes are dumped haphazardly.

## Environmental Issues

- Lack of proper collection and management.

- The 'throw away culture' result in these bags finding their way in the city drainage system and thus choking the drains.
- Littering of land by plastic bags presents an ugly and unhygienic scene.
- The littering also reduces rate of rain water percolation resulting in lowering water table levels.
- Plastics go into the water bodies which are already polluted due to many sources. Fish and other aquatic animals swallow plastic garbage mistaken as food items.
- Plastics become a nuisance because of their non-biodegradability.
- Animals eating carry bags sometimes die.
- Soil fertility deteriorates as plastic bags form part of manure and remain in soil for years.
- Polythene bags if burnt release highly toxic gases like phosgene, carbon monoxide, chlorine, sulphur dioxide, nitrogen oxide beside deadly dioxins.
- Requires large are for disposal and there are further waste disposal impactsrelated to land fills and incineration.

## Health Hazards

During the manufactureing process of polythene carry bags various harmful components/chemicals like colourants, pigments, plasticizers, antioxidants, stabilizers and heavy metals are used. Colours used during the process are mostly non food grade and leach out with other chemicals/ components thus contaminating food and other items carried in these bags. These chemicals can cause diseases like cancer, degeneration of brain tissues, heart enlargement etc.

Most of us take following metals with ready-to-eat food items, if carried in the cloured polythene carry bags:

- Lead through Black polythene carry bags.
- Chromium through Red polythene carry bags.
- Copper through Blue poythene carry bags.
- Salt of Barium through Green polythene carry bags.

## Plastic Pollution

Plastic pollution is currently one of the biggest environmental concerns. It may seem like large amounts of plastic waste are inevitable in the world we live in, but you can help with the plastic pollution issue by being aware of its dangers and taking steps to reduce waste.

The amount of garbage in the world increases as the population grows, and disposable plastic products, like water bottles and soda cans, accumulate over time. Plastic pollution occurs when enough plastic has gathered in an area that it affects the natural environment and harms plants, animals, or humans.

Plastic has toxic pollutants that damage the environment and cause land, water, and air pollution. It can take hundreds or even thousands of years for plastic to break down, so the damage to the environment is long-lasting.

## Cause of Plastic Pollution

Overuse of plastic is the main cause of plastic pollution. Plastic is cheap and widely available, but people frequently dispose of plastic items. They don't decompose, and they release an incredible amount of toxins into the air if they're burned.

Regular, everyday trash is one of the biggest contributors to plastic pollution. Milk cartons with plastic linings, disposable water bottles, soaps with small plastic beads, and other products end up in the environment or in dumps where they can affect the groundwater and nearby wildlife.

Commercial fishing nets are another big issue. Although fishing is necessary for the economy and for food supply in many regions, nets are often made of plastic. When the nets are submerged in the water, they leak toxins. They can also break or get lost, adding even more pollutants to the water.

## Effects of Plastic Pollution

The pollution in the ocean is mostly from plastic, and it has a terrible impact on marine species. As a result, it can hurt the economy and food supply for communities that rely on fishing.

Plastic can hurt tiny organisms like plankton, which larger animals rely on for food. If small organisms are poisoned from ingesting plastic, the animals that eat them will also consume toxins. The toxins work their way up the food chain and can even be present in the fish people eat.

Not only does plastic cause damage to the ocean, but it can also damage groundwater sources. Many regions are already facing issues with, but water sources everywhere are in danger because of plastic pollution. Plastic toxins in dumps and from litter can seep into the groundwater, which people drink every day.

On land, wind can carry plastic waste or litter throughout the environment. It can get stuck in trees, fences, traffic lights, or other structures. When animals come into contact with this plastic waste, they risk consuming the toxins or becoming entangled in the plastic and suffocating. If an animal consumes a piece of plastic, the plastic can clog its stomach while also poisoning it with toxins. Almost 200 different species of animals are known to ingest plastic debris.

Air pollution is another issue for humans and animals. When plastic is burned in the open air, it releases large amounts of toxins, which pollutes the air. If the toxins are inhaled for a long period of time, it can lead to respiratory problems.

As the world's population increases, land becomes more valuable, and it will soon become difficult to find places to put garbage. Over time, landfills and dumps will take up more land, invading animals' habitats and coming even closer to groundwater sources.

In addition to harming plants, animals, and people, it costs millions of dollars every year for clean-up of areas exposed to plastic toxins. Many regions have seen a decrease in tourism because of the amount of pollution in their environment, which can have a serious impact on local economies.

## Metal

Metal and tin waste includes scrap metal, broken metal cupboards/furniture, lead pots, equipment from which environmentally harmful substances such as asbestos and mercury have been removed (thermometers/manometers), oil, cooling motors, etc.

## Disposal

Metal waste from the buildings (workshops) can be collected by the supplier by arrangement. If large amounts of this waste are generated, it is advisable to arrange for a container to be provided. Removal of metal waste must be requested through FacilityNet.

## Processing

Tin cans are taken to the processing plants along with residual waste. At the processing plants they are separated from the other waste using magnets (for steel) or eddy currents (for aluminium). They are then smelted and used to produce new steel or aluminium. This is used to manufacture new products. New uses are always found for recycled tin: in bicycles, trains, bridges or as new cans.

Larger volumes of metal are taken to the metal processing plant where, in principle, the metal is shredded and reused in the same way.

## Heavy Metals Toxicity and the Environment

Heavy metals are naturally occurring elements that have a high atomic weight and a density at least 5 times greater than that of water. Their multiple industrial, domestic, agricultural, medical and technological applications have led to their wide distribution in the environment; raising concerns over their potential effects on human health and the environment. Their toxicity depends on several factors including the dose, route of exposure, and chemical species, as well as the age, gender, genetics, and nutritional status of exposed individuals. Because of their high degree of toxicity, arsenic, cadmium, chromium, lead, and mercury rank among the priority metals that are of public health significance. These metallic elements are considered systemic toxicants that are known to induce multiple organ damage, even at lower levels of exposure.

Heavy metals are defined as metallic elements that have a relatively high density compared to water. With the assumption that heaviness and toxicity are inter-related, heavy metals also include metalloids, such as arsenic, that are able to induce toxicity at low level of exposure. In recent years, there has been an increasing ecological and global public health concern associated with environmental contamination by these metals. Also, human exposure has risen dramatically as a result of an exponential increase of their use in several industrial, agricultural, domestic and technological applications. Reported sources of heavy metals in the environment include geogenic, industrial, agricultural, pharmaceutical, domestic effluents, and atmospheric sources. Environmental pollution is very prominent in point source areas such as mining, foundries and smelters, and other metal-based industrial operations.

Although heavy metals are naturally occurring elements that are found throughout the earth's crust, most environmental contamination and human exposure result from anthropogenic activities such as mining and smelting operations, industrial production and use, and domestic and agricultural use of metals and metal-containing compounds. Environmental contamination can also occur through metal corrosion, atmospheric deposition, soil erosion of metal ions and leaching of heavy metals, sediment re-suspension and metal evaporation from water resources to soil and ground water. Natural phenomena such as weathering and volcanic eruptions have also been reported to significantly contribute to heavy metal pollution. Industrial sources include metal processing in refineries, coal burning in power plants, petroleum combustion, nuclear power stations and high tension lines, plastics, textiles, microelectronics, wood preservation and paper processing plants.

It has been reported that metals such as cobalt (Co), copper (Cu), chromium (Cr), iron (Fe), magnesium (Mg), manganese (Mn), molybdenum (Mo), nickel (Ni), selenium (Se) and zinc (Zn) are essential nutrients that are required for various biochemical and physiological functions. Inadequate supply of these micro-nutrients results in a variety of deficiency diseases or syndromes.

Heavy metals are also considered as trace elements because of their presence in trace concentrations (ppb range to less than 10ppm) in various environmental matrices. Their bioavailability is influenced by physical factors such as temperature, phase association, adsorption and sequestration. It is also affected by chemical factors that influence speciation at thermodynamic equilibrium, complexation kinetics, lipid solubility and octanol/water partition coefficients. Biological factors such as species characteristics, trophic interactions, and biochemical/physiological adaptation, also play an important role.

The essential heavy metals exert biochemical and physiological functions in plants and animals. They are important constituents of several key enzymes and play important roles in various oxidation-reduction reactions. Copper for example serves as an essential co-factor for several oxidative stress-related enzymes including catalase, superoxide dismutase, peroxidase, cytochrome c oxidases, ferroxidases, monoamine oxidase, and dopamine β-monooxygenase. Hence, it is an essential nutrient that is incorporated into a number of metalloenzymes involved in hemoglobin formation, carbohydrate metabolism, catecholamine biosynthesis, and cross-linking of collagen, elastin, and hair keratin. The ability of copper to cycle between an oxidized state, Cu(II), and reduced state, Cu(I), is used by cuproenzymes involved in redox reactions. However, it is this property of copper that also makes it potentially toxic because the transitions between Cu(II) and Cu(I) can result in the generation of superoxide and hydroxyl radicals. Also, excessive exposure to copper has been linked to cellular damage leading to Wilson disease in humans. Similar to copper, several other essential elements are required for biologic functioning, however, an excess amount of such metals produces cellular and tissue damage leading to a variety of adverse effects and human diseases. For some including chromium and copper, there is a very narrow range of concentrations between beneficial and toxic effects. Other metals such as aluminium (Al), antinomy (Sb), arsenic (As), barium (Ba), beryllium (Be), bismuth (Bi), cadmium (Cd), gallium (Ga), germanium (Ge), gold (Au), indium (In), lead (Pb), lithium (Li), mercury (Hg), nickel (Ni), platinum (Pt), silver (Ag), strontium (Sr), tellurium (Te), thallium (Tl), tin (Sn), titanium (Ti), vanadium (V) and uranium (U) have no established biological functions and are considered as non-essential metals.

In biological systems, heavy metals have been reported to affect cellular organelles and components such as cell membrane, mitochondrial, lysosome, endoplasmic reticulum, nuclei, and some enzymes involved in metabolism, detoxification, and damage repair. Metal ions have been found to interact with cell components such as DNA and nuclear proteins, causing DNA damage and conformational changes that may lead to cell cycle modulation, carcinogenesis or apoptosis. Several studies from our laboratory have demonstrated that reactive oxygen species (ROS) production and oxidative stress play a key role in the toxicity and carcinogenicity of metals such as arsenic, cadmium, chromium, lead, and mercury. Because of their high degree of toxicity, these five elements rank among the priority metals that are of great public health significance. They are all systemic toxicants that are known to induce multiple organ damage, even at lower levels of exposure. According to the United States Environmental Protection Agency (U.S. EPA), and the International Agency for Research on Cancer (IARC), these metals are also classified as either "known" or "probable" human carcinogens based on epidemiological and experimental studies showing an association between exposure and cancer incidence in humans and animals.

Heavy metal-induced toxicity and carcinogenicity involves many mechanistic aspects, some of which are not clearly elucidated or understood. However, each metal is known to have unique features and physic-chemical properties that confer to its specific toxicological mechanisms of action. Here we analyse the environmental occurrence, production and use, potential for human exposure, and molecular mechanisms of toxicity, genotoxicity, and carcinogenicity of arsenic, cadmium, chromium, lead, and mercury.

## Arsenic

### Environmental Occurrence, Industrial Production and Use

Arsenic is a ubiquitous element that is detected at low concentrations in virtually all environmental matrices. The major inorganic forms of arsenic include the trivalent arsenite and the pentavalent arsenate. The organic forms are the methylated metabolites – monomethylarsonic acid (MMA), dimethylarsinic acid (DMA) and trimethylarsine oxide. Environmental pollution by arsenic occurs as a result of natural phenomena such as volcanic eruptions and soil erosion, and anthropogenic activities. Several arsenic-containing compounds are produced industrially, and have been used to manufacture products with agricultural applications such as insecticides, herbicides, fungicides, algicides, sheep dips, wood preservatives, and dye-stuffs. They have also been used in veterinary medicine for the eradication of tapeworms in sheep and cattle. Arsenic compounds have also been used in the medical field for at least a century in the treatment of syphilis, yaws, amoebic dysentery, and trypanosomaiasis. Arsenic-based drugs are still used in treating certain tropical diseases such as African sleeping sickness and amoebic dysentery, and in veterinary medicine to treat parasitic diseases, including filariasis in dogs and black head in turkeys and chickens. Recently, arsenic trioxide has been approved by the Food and Drug Administration as an anticancer agent in the treatment of acute promeylocytic leukemia. Its therapeutic action has been attributed to the induction of programmed cell death (apoptosis) in leukemia cells.

## Potential for Human Exposure

It is estimated that several million people are exposed to arsenic chronically throughout the world, especially in countries like Bangladesh, India, Chile, Uruguay, Mexico, Taiwan, where the ground water is contaminated with high concentrations of arsenic. Exposure to arsenic occurs via the oral route (ingestion), inhalation, dermal contact, and the parenteral route to some extent. Arsenic concentrations in air range from 1 to 3 ng/m$^3$ in remote locations (away from human releases), and from 20 to 100 ng/m$^3$ in cities. Its water concentration is usually less than 10μg/L, although higher levels can occur near natural mineral deposits or mining sites. Its concentration in various foods ranges from 20 to 140 ng/kg. Natural levels of arsenic in soil usually range from 1 to 40 mg/kg, but pesticide application or waste disposal can produce much higher values.

Diet, for most individuals, is the largest source of exposure, with an average intake of about 50 μg per day. Intake from air, water and soil are usually much smaller, but exposure from these media may become significant in areas of arsenic contamination. Workers who produce or use arsenic compounds in such occupations as vineyards, ceramics, glass-making, smelting, refining of metallic ores, pesticide manufacturing and application, wood preservation, semiconductor manufacturing can be exposed to substantially higher levels of arsenic. Arsenic has also been identified at 781 sites of the 1,300 hazardous waste sites that have been proposed by the U.S. EPA for inclusion on the national priority list. Human exposure at these sites may occur by a variety of pathways, including inhalation of dusts in air, ingestion of contaminated water or soil, or through the food chain.

Contamination with high levels of arsenic is of concern because arsenic can cause a number of human health effects. Several epidemiological studies have reported a strong association between arsenic exposure and increased risks of both carcinogenic and systemic health effects. Interest in the toxicity of arsenic has been heightened by recent reports of large populations in West Bengal, Bangladesh, Thailand, Inner Mongolia, Taiwan, China, Mexico, Argentina, Chile, Finland and Hungary that have been exposed to high concentrations of arsenic in their drinking water and are displaying various clinico-pathological conditions including cardiovascular and peripheral vascular disease, developmental anomalies, neurologic and neurobehavioural disorders, diabetes, hearing loss, portal fibrosis, hematologic disorders (anemia, leukopenia and eosinophilia) and carcinoma. Arsenic exposure affects virtually all organ systems including the cardiovascular, dermatologic, nervous, hepatobilliary, renal, gastro-intestinal, and respiratory systems. Research has also pointed to significantly higher standardized mortality rates for cancers of the bladder, kidney, skin, and liver in many areas of arsenic pollution. The severity of adverse health effects is related to the chemical form of arsenic, and is also time-and dose-dependent. Although the evidence of carcinogenicity of arsenic in humans seems strong, the mechanism by which it produces tumors in humans is not completely understood.

## Mechanisms of Toxicity and Carcinogenicity

Analyzing the toxic effects of arsenic is complicated because the toxicity is highly influenced by its oxidation state and solubility, as well as many other intrinsic and extrinsic factors. Several studies have indicated that the toxicity of arsenic depends on the exposure dose, frequency and duration, the biological species, age, and gender, as well as on individual susceptibilities, genetic and nutritional factors. Most cases of human toxicity from arsenic have been associated with exposure to inorganic arsenic. Inorganic trivalent arsenite ($As^{III}$) is 2–10 times more toxic than pentavalent arsenate ($As^{V}$). By binding to thiol or sulfhydryl groups on proteins, As (III) can inactivate over 200 enzymes. This is the likely mechanism responsible for arsenic's widespread effects on different organ systems. As (V) can replace phosphate, which is involved in many biochemical pathways.

One of the mechanisms by which arsenic exerts its toxic effect is through impairment of cellular respiration by the inhibition of various mitochondrial enzymes, and the uncoupling of oxidative phosphorylation. Most toxicity of arsenic results from its ability to interact with sulfhydryl groups of proteins and enzymes, and to substitute phosphorous in a variety of biochemical reactions. Arsenic *in vitro* reacts with protein sulfhydryl groups to inactivate enzymes, such as dihydrolipoyl dehydrogenase and thiolase, thereby producing inhibited oxidation of pyruvate and betaoxidation of fatty acids. The major metabolic pathway for inorganic arsenic in humans is methylation. Arsenic trioxide is methylated to two major metabolites via a non-enzymatic process to monomethylarsonic acid (MMA), which is further methylated enzymatically to dimethyl arsenic acid (DMA) before excretion in the urine. It was previously thought that this methylation process is a pathway of arsenic detoxification, however, recent studies have pointed out that some methylated metabolites may be more toxic than arsenite if they contain trivalent forms of arsenic.

Tests for genotoxicity have indicated that arsenic compounds inhibit DNA repair, and induce chromosomal aberrations, sister-chromatid exchanges, and micronuclei formation in both human and rodent cells in culture and in cells of exposed humans. Reversion assays with *Salmonella typhimurium* fail to detect mutations that are induced by arsenic compounds. Although arsenic compounds are generally perceived as weak mutagens in bacterial and animal cells, they exhibit clastogenic properties in many cell types *in vivo* and *in vitro*. In the absence of animal models, *in vitro* cell transformation studies become a useful means of obtaining information on the carcinogenic mechanisms of arsenic toxicity. Arsenic and arsenical compounds are cytotoxic and induce morphological transformations of Syrian hamster embryo (SHE) cells as well as mouse C3H10T1/2 cells and BALB/3T3 cells.

Based on the comet assay, it has been reported that arsenic trioxide induces DNA damage in human lymphophytes and also in mice leukocytes. Arsenic compounds have also been shown to induce gene amplification, arrest cells in mitosis, inhibit DNA repair, and induce expression of the *c-fos* gene and the oxidative stress protein heme oxygenase in mammalian cells. They have been implicated as promoters and comutagens for a variety of toxic agents. Epidemiological investigations have indicated that long-term arsenic exposure results in promotion of carcinogenesis. Several hypotheses have been proposed to describe the mechanism of arsenic-induced carcinogenesis. Zhao et al. reported that arsenic may act as a carcinogen by inducing DNA hypomethylation, which in turn facilitates aberrant gene expression. Additionally, it was found that arsenic is a potent stimulator of extracellular signal-regulated protein kinase Erk1 and AP-1 transactivational activity, and an efficient inducer of *c-fos* and *c-jun* gene expression. Induction of *c-jun* and *c-fos* by arsenic is associated with activation of JNK. However, the role of JNK activation by arsenite in cell transformation or tumor promotion is unclear.

Trouba et al. concluded that long-term exposure to high levels of arsenic might make cells more susceptible to mitogenic stimulation and that alterations in mitogenic signaling proteins might contribute to the carcinogenic action of arsenic. Collectively, several recent studies have demonstrated that arsenic can interfere with cell signaling pathways (e.g., the p53 signaling pathway) that are frequently implicated in the promotion and progression of a variety of tumor types in experimental animal models, and of some human tumors. However, the specific alterations in signal transduction pathways or the actual targets that contribute to the development of arsenic-induced tumors in humans following chronic consumption of arsenic remains uncertain.

Recent clinical trials have found that arsenic trioxide has therapeutic value in the treatment of acute promyelocytic leukemia, and there is interest in exploring its effectiveness in the treatment of a variety of other cancers. In acute promyelocytic leukemia, the specific molecular event critical to the formation of malignant cells is known. A study by Puccetti et al. found that forced overexpression of BCR-ABL susceptibility in human lymphoblasts cells resulted in greatly enhanced sensitivity to arsenic-induced apoptosis. They also concluded that arsenic trioxide is a tumor specific agent capable of inducing apoptosis selectively in acute promyelocytic leukemia cells. Several recent studies have shown that arsenic can induce apoptosis through alterations in other cell signaling pathways. In addition to acute peomyelocytic leukemia, arsenic is thought to have therapeutic potential for myeloma. In summary, numerous cancer chemotherapy studies in cell cultures and in patients with acute promyelocytic leukemia demonstrate that arsenic trioxide administration can lead to cell-cycle arrest and apoptosis in malignant cells.

Previous studies have also examined p53 gene expression and mutation in tumors obtained from subjects with a history of arsenic ingestion. p53 participates in many cellular functions, cell-cycle control, DNA repair, differentiation, genomic plasticity and programmed cell death. Additional support for the hypothesis that arsenic can modulate gene expression has been provided by several different studies. Collectively, these studies provide further evidence that various forms of arsenic can alter gene expression and that such changes could contribute substantially to the toxic and carcinogenic actions of arsenic treatment in human populations.

## Cadmium

### Environmental Occurrence, Industrial Production and Use

Cadmium is a heavy metal of considerable environmental and occupational concern. It is widely distributed in the earth's crust at an average concentration of about 0.1 mg/kg. The highest level of cadmium compounds in the environment is accumulated in sedimentary rocks, and marine phosphates contain about 15 mg cadmium/kg.

Cadmium is frequently used in various industrial activities. The major industrial applications of cadmium include the production of alloys, pigments, and batteries. Although the use of cadmium in batteries has shown considerable growth in recent years, its commercial use has declined in developed countries in response to environmental concerns. In the United States for example, the daily cadmium intake is about 0.4μg/kg/day, less than half of the U.S. EPA's oral reference dose.

This decline has been linked to the introduction of stringent effluent limits from plating works and, more recently, to the introduction of general restrictions on cadmium consumption in certain countries.

## Potential for Human Exposure

The main routes of exposure to cadmium are via inhalation or cigarette smoke, and ingestion of food. Skin absorption is rare. Human exposure to cadmium is possible through a number of several sources including employment in primary metal industries, eating contaminated food, smoking cigarettes, and working in cadmium-contaminated work places, with smoking being a major contributor. Other sources of cadmium include emissions from industrial activities, including mining, smelting, and manufacturing of batteries, pigments, stabilizers, and alloys. Cadmium is also present in trace amounts in certain foods such as leafy vegetables, potatoes, grains and seeds, liver and kidney, and crustaceans and mollusks. In addition, foodstuffs that are rich in cadmium can greatly increase the cadmium concentration in human bodies. Examples are liver, mushrooms, shellfish, mussels, cocoa powder and dried seaweed. An important distribution route is the circulatory system whereas blood vessels are considered to be main stream organs of cadmium toxicity. Chronic inhalation exposure to cadmium particulates is generally associated with changes in pulmonary function and chest radiographs that are consistent with emphysema. Workplace exposure to airborne cadmium particulates has been associated with decreases in olfactory function. Several epidemiologic studies have documented an association of chronic low-level cadmium exposure with decreases in bone mineral density and osteoporosis.

Exposure to cadmium is commonly determined by measuring cadmium levels in blood or urine. Blood cadmium reflects recent cadmium exposure (from smoking, for example). Cadmium in urine (usually adjusted for dilution by calculating the cadmium/creatinine ratio) indicates accumulation, or kidney burden of cadmium. It is estimated that about 2.3% of the U.S. population has elevated levels of urine cadmium (>2μg/g creatinine), a marker of chronic exposure and body burden. Blood and urine cadmium levels are typically higher in cigarette smokers, intermediate in former smokers and lower in nonsmokers. Because of continuing use of cadmium in industrial applications, the environmental contamination and human exposure to cadmium have dramatically increased during the past century.

## Molecular Mechanisms of Toxicity and Carcinogenicity

Cadmium is a severe pulmonary and gastrointestinal irritant, which can be fatal if inhaled or ingested. After acute ingestion, symptoms such as abdominal pain, burning sensation, nausea, vomiting, salivation, muscle cramps, vertigo, shock, loss of consciousness and convulsions usually appear within 15 to 30 min. Acute cadmium ingestion can also cause gastrointestinal tract erosion, pulmonary, hepatic or renal injury and coma, depending on the route of poisoning. Chronic exposure to cadmium has a depressive effect on levels of norepinephrine, serotonin, and acetylcholine. Rodent studies have shown that chronic inhalation of cadmium causes pulmonary adenocarcinomas. It can also cause prostatic proliferative lesions including adenocarcinomas, after systemic or direct exposure.

Although the mechanisms of cadmium toxicity are poorly understood, it has been speculated that cadmium causes damage to cells primarily through the generation of ROS, which causes single-strand DNA damage and disrupts the synthesis of nucleic acids and proteins. Studies using two-dimensional gel electrophoresis have shown that several stress response systems are expressed in response to cadmium exposure, including those for heat shock, oxidative stress, stringent response, cold shock, and SOS. *In vitro* studies indicate that cadmium induces cytotoxic effects at the concentrations 0.1 to 10 mM and free radical-dependent DNA damage. *In vivo* studies have shown that cadmium modulates male reproduction in mice model at a concentration of 1 mg/kg body weight. However, cadmium is a weak mutagen when compared with other carcinogenic metals. Previous reports have indicated that cadmium affects signal transduction pathways; inducing inositol polyphosphate formation, increasing cytosolic free calcium levels in various cell types, and blocking calcium channels. At lower concentrations (1–100 μM), cadmium binds to proteins, decreases DNA repair, activates protein degradation, up-regulates cytokines and proto-oncogenes such as c-*fos*, c-*jun*, and c-*myc* , and induces expression of several genes including metallothioneins, heme oxygenases, glutathione transferases, heat-shock proteins, acute-phase reactants, and DNA polymerase β.

Cadmium compounds are classified as human carcinogens by several regulatory agencies. The International Agency for Research on Cancer and the U.S. National Toxicology Program have concluded that there is adequate evidence that cadmium is a human carcinogen. This designation as a human carcinogen is based primarily on repeated findings of an association between occupational cadmium exposure and lung cancer, as well as on very strong rodent data showing the pulmonary system as a target site. Thus, the lung is the most definitively established site of human carcinogenesis from cadmium exposure. Other target tissues of cadmium carcinogenesis in animals include injection sites, adrenals, testes, and the hemopoietic system. In some studies, occupational or environmental cadmium exposure has also been associated with development of cancers of the prostate, kidney, liver, hematopoietic system and stomach. Carcinogenic metals including arsenic, cadmium, chromium, and nickel have all been associated with DNA damage through base pair mutation, deletion, or oxygen radical attack on DNA. Animal studies have demonstrated reproductive and teratogenic effects. Small epidemiologic studies have noted an inverse relationship between cadmium in cord blood, maternal blood or maternal urine and birth weight and length at birth.

## Chromium

### Environmental Occurrence, Industrial Production and Use

Chromium (Cr) is a naturally occurring element present in the earth's crust, with oxidation states (or valence states) ranging from chromium (II) to chromium (VI). Chromium compounds are stable in the trivalent [Cr(III)] form and occur in nature in this state in ores, such as ferrochromite. The hexavalent [Cr(VI)] form is the second-most stable state. Elemental chromium [Cr(0)] does not occur naturally. Chromium enters into various environmental matrices (air, water, and soil) from a wide variety of natural and anthropogenic sources with the largest release coming from industrial establishments. Industries with the largest contribution to chromium release include metal processing, tannery facilities, chromate production, stainless steel welding, and ferrochrome and chrome pigment production. The increase in the environmental concentrations of chromium has been linked to air and wastewater release of chromium, mainly from metallurgical, refractory, and chemical industries. Chromium released into the environment from anthropogenic activity occurs mainly in the hexavalent form [Cr(VI)]. Hexavalent chromium [Cr(VI)] is a toxic industrial pollutant that is classified as human carcinogen by several regulatory and non-regulatory agencies. The health hazard associated with exposure to chromium depends on its oxidation state, ranging from the low toxicity of the metal form to the high toxicity of the hexavalent form. All Cr(VI)-containing compounds were once thought to be man-made, with only Cr(III) naturally ubiquitous in air, water, soil and biological materials. Recently, however, naturally occurring Cr(VI) has been found in ground and surface waters at values exceeding the World Health Organization limit for drinking water of 50 μg of Cr(VI) per liter. Chromium is widely used in numerous industrial processes and as a result, is a contaminant of many environmental systems. Commercially chromium compounds are used in industrial welding, chrome plating, dyes and pigments, leather tanning and wood preservation. Chromium is also used as anticorrosive in cooking systems and boilers.

## Potential for Human Exposure

It is estimated that more than 300,000 workers are exposed annually to chromium and chromium-containing compounds in the workplace. In humans and animals, [Cr(III)] is an essential nutrient that plays a role in glucose, fat and protein metabolism by potentiating the action of insulin. However, occupational exposure has been a major concern because of the high risk of Cr-induced diseases in industrial workers occupationally exposed to Cr(VI). Also, the general human population and some wildlife may also be at risk. It is estimated that 33 tons of total Cr are released annually into the environment. The U.S. Occupational Safety and Health Administration (OSHA) recently set a "safe" level of 5μg/m$^3$, for an 8-hr time-weighted average, even though this revised level may still pose a carcinogenic risk. For the general human population, atmospheric levels range from 1 to 100 ng/cm$^3$, but can exceed this range in areas that are close to Cr manufacturing.

Non-occupational exposure occurs via ingestion of chromium containing food and water whereas occupational exposure occurs via inhalation. Chromium concentrations range between 1 and 3000 mg/kg in soil, 5 to 800 μg/L in sea water, and 26 μg/L to 5.2 mg/L in rivers and lakes. Chromium content in foods varies greatly and depends on the processing and preparation. In general, most fresh foods typically contain chromium levels ranging from <10 to 1,300 μg/kg. Present day workers in chromium-related industries can be exposed to chromium concentrations two orders of magnitude higher than the general population. Even though the principal route of human exposure to chromium is through inhalation, and the lung is the primary target organ, significant human exposure to chromium has also been reported to take place through the skin. For example, the widespread incidence of dermatitis noticed among construction workers is attributed to their exposure to chromium present in cement. Occupational and environmental exposure to Cr(VI)-containing compounds is known to cause multiorgan toxicity such as renal damage, allergy and asthma, and cancer of the respiratory tract in humans.

Breathing high levels of chromium (VI) can cause irritation to the lining of the nose, and nose ulcers. The main health problems seen in animals following ingestion of chromium (VI) compounds are irritation and ulcers in the stomach and small intestine, anemia, sperm damage and male reproductive system damage. Chromium (III) compounds are much less toxic and do not appear to cause these problems. Some individuals are extremely sensitive to chromium(VI) or chromium(III), allergic reactions consisting of severe redness and swelling of the skin have been noted. An increase in stomach tumors was observed in humans and animals exposed to chromium(VI) in drinking water. Accidental or intentional ingestion of extremely high doses of chromium (VI) compounds by humans has resulted in severe respiratory, cardiovascular, gastrointestinal, hematological, hepatic, renal, and neurological effects as part of the sequelae leading to death or in patients who survived because of medical treatment. Although the evidence of carcinogenicity of chromium in humans and terrestrial mammals seems strong, the mechanism by which it causes cancer is not completely understood.

## Mechanisms of Toxicity and Carcinogenicity

Major factors governing the toxicity of chromium compounds are oxidation state and solubility. Cr(VI) compounds, which are powerful oxidizing agents and thus tend to be irritating and corrosive, appear to be much more toxic systemically than Cr(III) compounds, given similar amount and solubility. Although the mechanisms of biological interaction are uncertain, the variation in toxicity may be related to the ease with which Cr(VI) can pass through cell membranes and its subsequent intracellular reduction to reactive intermediates. Since Cr(III) is poorly absorbed by any route, the toxicity of chromium is mainly attributable to the Cr(VI) form. It can be absorbed by the lung and gastrointestinal tract, and even to a certain extent by intact skin. The reduction of Cr(VI) is considered as being a detoxification process when it occurs at a distance from the target site for toxic or genotoxic effect while reduction of Cr(VI) may serve to activate chromium toxicity if it takes place in or near the cell nucleus of target organs. If Cr(VI) is reduced to Cr(III) extracellularly, this form of the metal is not readily transported into cells and so toxicity is not observed. The balance that exists between extracellular Cr(VI) and intracellular Cr(III) is what ultimately dictates the amount and rate at which Cr(VI) can enter cells and impart its toxic effects.

Cr(VI) enters many types of cells and under physiological conditions can be reduced by hydrogen peroxide ($H_2O_2$), glutathione (GSH) reductase, ascorbic acid, and GSH to produce reactive intermediates, including Cr(V), Cr(IV), thiylradicals, hydroxyl radicals, and ultimately, Cr(III). Any of these species could attack DNA, proteins, and membrane lipids, thereby disrupting cellular integrity and functions.

Studies with animal models have also reported many harmful effects of Cr (VI) on mammals. Subcutaneous administration of Cr (VI) to rats caused severe progressive proteinuria, urea nitrogen and creatinine, as well as elevation in serum alanine aminotransferase activity and hepatic lipid peroxide formation. Similar studies reported by Gumbleton and Nicholls found that Cr (VI) induced renal damage in rats when administered by single sub-cutaneous injections. Bagchi et al. demonstrated that rats received Cr (VI) orally in water induced hepatic mitochondrial and microsomal lipid peroxidation, as well as enhanced excretion of urinary lipid metabolites including malondialdehyde.

Adverse health effects induced by Cr (VI) have also been reported in humans. Epidemiological investigations have reported respiratory cancers in workers occupationally exposed to Cr (VI)-containing compounds. DNA strand breaks in peripheral lymphocytes and lipid peroxidation products in urine observed in chromium-exposed workers also support the evidence of Cr (VI)-induced toxicity to humans. Oxidative damage is considered to be the underlying cause of these genotoxic effects including chromosomal abnormalities, and DNA strand breaks. Nevertheless, recent studies indicate a biological relevance of non-oxidative mechanisms in Cr(VI) carcinogenesis.

Carcinogenicity appears to be associated with the inhalation of the less soluble/insoluble Cr(VI) compounds. The toxicology of Cr(VI) does not reside with the elemental form. It varies greatly among a wide variety of very different Cr(VI) compounds. Epidemiological evidence strongly points to Cr(VI) as the agent in carcinogenesis. Solubility and other characteristics of chromium, such as size, crystal modification, surface charge, and the ability to be phagocytized might be important in determining cancer risk.

## Lead

### Environmental Occurrence, Industrial Production and Use

Lead is a naturally occurring bluish-gray metal present in small amounts in the earth's crust. Although lead occurs naturally in the environment, anthropogenic activities such as fossil fuels burning, mining, and manufacturing contribute to the release of high concentrations. Lead has many different industrial, agricultural and domestic applications. It is currently used in the production of lead-acid batteries, ammunitions, metal products (solder and pipes), and devices to shield X-rays. An estimated 1.52 million metric tons of lead were used for various industrial applications in the United Stated in 2004. Of that amount, lead-acid batteries production accounted for 83 percent, and the remaining usage covered a range of products such as ammunitions (3.5 percent), oxides for paint, glass, pigments and chemicals (2.6 percent), and sheet lead (1.7 percent).

In recent years, the industrial use of lead has been significantly reduced from paints and ceramic products, caulking, and pipe solder. Despite this progress, it has been reported that among 16.4 million United States homes with more than one child younger than 6 years per household, 25% of homes still had significant amounts of lead-contaminated deteriorated paint, dust, or adjacent bare soil. Lead in dust and soil often re-contaminates cleaned houses and contributes to elevating blood lead concentrations in children who play on bare, contaminated soil. Today, the largest source of lead poisoning in children comes from dust and chips from deteriorating lead paint on interior surfaces. Children who live in homes with deteriorating lead paint can achieve blood lead concentrations of 20μg/dL or greater.

### Potential for Human Exposure

Exposure to lead occurs mainly via inhalation of lead-contaminated dust particles or aerosols, and ingestion of lead-contaminated food, water, and paints. Adults absorb 35 to 50% of lead through drinking water and the absorption rate for children may be greater than 50%. Lead absorption is influenced by factors such as age and physiological status. In the human body, the greatest percentage of lead is taken into the kidney, followed by the liver and the other soft tissues such as heart and brain, however, the lead in the skeleton represents the major body fraction. The nervous system is the most vulnerable target of lead poisoning. Headache, poor attention spam, irritability, loss of memory and dullness are the early symptoms of the effects of lead exposure on the central nervous system.

Since the late 1970's, lead exposure has decreased significantly as a result of multiple efforts including the elimination of lead in gasoline, and the reduction of lead levels in residential paints, food and drink cans, and plumbing systems. Several federal programs implemented by state and local health governments have not only focused on banning lead in gasoline, paint and soldered cans, but have also supported screening programs for lead poisoning in children and lead abatement in housing. Despite the progress in these programs, human exposure to lead remains a serious health problem. Lead is the most systemic toxicant that affects several organs in the body including the kidneys, liver, central nervous system, hematopoetic system, endocrine system, and reproductive system.

Lead exposure usually results from lead in deteriorating household paints, lead in the work place, lead in crystals and ceramic containers that leaches into water and food, lead use in hobbies, and lead use in some traditional medicines and cosmetics. Several studies conducted by the National Health and Nutrition Examination surveys (NHANES) have measured blood lead levels in the U.S. populations and have assessed the magnitude of lead exposure by age, gender, race, income and degree of urbanization. Although the results of these surveys have demonstrated a general decline in blood lead levels since the 1970s, they have also shown that large populations of children continue to have elevated blood lead levels (> 10μg/dL). Hence, lead poisoning remains one of the most common pediatric health problems in the United States today. Exposure to lead is of special concern among women particularly during pregnancy. Lead absorbed by the pregnant mother is readily transferred to the developing fetus. Human evidence corroborates animal findings, linking prenatal exposure to lead with reduced birth weight and preterm delivery, and with neuro-developmental abnormalities in offspring.

## Molecular Mechanisms of Toxicity and Carcinogenicity

There are many published studies that have documented the adverse effects of lead in children and the adult population. In children, these studies have shown an association between blood level poisoning and diminished intelligence, lower intelligence quotient-IQ, delayed or impaired neurobehavioral development, decreased hearing acuity, speech and language handicaps, growth retardation, poor attention span, and anti social and diligent behaviors. In the adult population, reproductive effects, such as decreased sperm count in men and spontaneous abortions in women have been associated with high lead exposure. Acute exposure to lead induces brain damage, kidney damage, and gastrointestinal diseases, while chronic exposure may cause adverse effects on the blood, central nervous system, blood pressure, kidneys, and vitamin D metabolism.

One of the major mechanisms by which lead exerts its toxic effect is through biochemical processes that include lead's ability to inhibit or mimic the actions of calcium and to interact with proteins. Within the skeleton, lead is incorporated into the mineral in place of calcium. Lead binds to biological molecules and thereby interfering with their function by a number of mechanisms. Lead binds to sulfhydryl and amide groups of enzymes, altering their configuration and diminishing their activities. Lead may also compete with essential metallic cations for binding sites, inhibiting enzyme activity, or altering the transport of essential cations such as calcium. Many investigators have demonstrated that lead intoxication induces a cellular damage mediated by the formation of reactive oxygen species (ROS). In addition, Jiun and Hseien demonstrated that the levels of malondialdehyde (MDA) in blood strongly correlate with lead concentration in the blood of exposed workers. Other studies showed that the activities of antioxidant enzymes, including superoxide dismutase (SOD), and glutathione peroxidase in erythrocytes of workers exposed to lead are remarkably higher than that in non-exposed workers. Lead-induced toxicity and apoptosis in human cancer cells involved several cellular and molecular processes including induction of cell death and oxidative stress, transcriptional activation of stress genes, DNA damage, externalization of phosphatidylserine and activation of caspase-3.

A large body of research has indicated that lead acts by interfering with calcium-dependent processes related to neuronal signaling and intracellular signal transduction. Lead perturbs intracellular calcium cycling, altering releasability of organelle stores, such as endoplasmic reticulum and mitochondria. In some cases lead inhibits calcium-dependent events, including calcium-dependent release of several neurotransmitters and receptor-coupled ionophores in glutamatergic neurons. In other cases lead appears to augment calcium-dependent events, such as protein kinase C and calmodulin.

Experimental studies have indicated that lead is potentially carcinogenic, inducing renal tumors in rats and mice, and is therefore considered by the IARC as a probable human carcinogen. Lead exposure is also known to induce gene mutations and sister chromatid exchanges, morphological transformations in cultured rodent cells, and to enhance anchorage independence in diploid human fibroblasts. *In vitro* and *in vivo* studies indicated that lead compounds cause genetic damage through various indirect mechanisms that include inhibition of DNA synthesis and repair, oxidative damage, and interaction with DNA-binding proteins and tumor suppressor proteins. Studies by Roy and his group showed that lead acetate induced mutagenicity at a toxic dose at the *E. coli gpt* locus transfected to V79 cells. They also reported that toxic doses of lead acetate and lead nitrate induced DNA breaks at the *E. coli gpt* locus transfected to V79 cells. Another study by Wise and his collaborators found no evidence for direct genotoxic or DNA-damaging effects of lead except for lead chromate. They pointed out that the genotoxicity may be due to hexavalent chromate rather than lead.

## Mercury

### Environmental Occurrence, Industrial Production and Use

Mercury is a heavy metal belonging to the transition element series of the periodic table. It is unique in that it exists or is found in nature in three forms (elemental, inorganic, and organic), with each having its own profile of toxicity. At room temperature elemental mercury exists as a liquid which has a high vapor pressure and is released into the environment as mercury vapor. Mercury also exists as a cation with oxidation states of +1 (mercurous) or +2 (mercuric). Methylmercury is the most frequently encountered compound of the organic form found in the environment, and is formed as a result of the methylation of inorganic (mercuric) forms of mercury by microorganisms found in soil and water.

Mercury is a widespread environmental toxicant and pollutant which induces severe alterations in the body tissues and causes a wide range of adverse health effects. Both humans and animals are exposed to various chemical forms of mercury in the environment. These include elemental mercury vapor ($Hg^0$), inorganic mercurous ($Hg^{+1}$), mercuric ($Hg^{+2}$), and the organic mercury compounds. Because mercury is ubiquitous in the environment, humans, plants and animals are all unable to avoid exposure to some form of mercury.

Mercury is utilized in the electrical industry (switches, thermostats, batteries), dentistry (dental amalgams), and numerous industrial processes including the production of caustic soda, in nuclear reactors, as antifungal agents for wood processing, as a solvent for reactive and precious metal, and as a preservative of pharmaceutical products. The industrial demand for mercury peaked in 1964 and began to sharply decline between 1980 and 1994 as a result of federal bans on mercury additives in paints, pesticides, and the reduction of its use in batteries.

## Potential for Human Exposure

Humans are exposed to all forms of mercury through accidents, environmental pollution, food contamination, dental care, preventive medical practices, industrial and agricultural operations, and occupational operations. The major sources of chronic, low level mercury exposure are dental amalgams and fish consumption. Mercury enters water as a natural process of off-gassing from the earth's crust and also through industrial pollution. Algae and bacteria methylate the mercury entering the waterways. Methyl mercury then makes its way through the food chain into fish, shell-fish, and eventually into humans.

The two most highly absorbed species are elemental mercury ($Hg^o$) and methyl mercury (MeHg). Dental amalgams contain over 50% elemental mercury. The elemental vapor is highly lipophilic and is effectively absorbed through the lungs and tissues lining the mouth. After $Hg^o$ enters the blood, it rapidly passes through cell membranes, which include both the blood-brain barrier and the placental barrier. Once it gains entry into the cell, $Hg^o$ is oxidized and becomes highly reactive $Hg^{2+}$. Methyl mercury derived from eating fish is readily absorbed in the gastrointestinal tract and because of its lipid solubility, can easily cross both the placental and blood-brain barriers. Once mercury is absorbed it has a very low excretion rate. A major proportion of what is absorbed accumulates in the kidneys, neurological tissue and the liver. All forms of mercury are toxic and their effects include gastrointestinal toxicity, neurotoxicity, and nephrotoxicity.

## Molecular Mechanisms of Mercury Toxicity and Carcingenicity

The molecular mechanisms of toxicity of mercury are based on its chemical activity and biological features which suggest that oxidative stress is involved in its toxicity. Through oxidative stress mercury has shown mechanisms of sulfhydryl reactivity. Once in the cell both $Hg^{2+}$ and MeHg form covalent bonds with cysteine residues of proteins and deplete cellular antioxidants. Antioxidant enzymes serve as a line of cellular defense against mercury compounds. The interaction of mercury compounds suggests the production of oxidative damage through the accumulation of reactive oxygen species (ROS) which would normally be eliminated by cellular antioxidants.

In eukaryotic organisms the primary site for the production of reactive oxygen species (ROS) occurs in the mitochondria through normal metabolism. Inorganic mercury has been reported to increase the production of these ROS by causing defects in oxidative phosphorylation and electron transport at the ubiquinone-cytochrome b5 step. Through the acceleration of the rate of electron transfer in the electron transport chain in the mitochondria, mercury induces the premature shedding of electrons to molecular oxygen which causes an increase in the generation of reactive oxygen species.

Oxidative stress appears to also have an effect on calcium homeostasis. The role of calcium in the activation of proteases, endonucleases and phospholipases is well established. The activation of phospholipase $A_2$ has been shown to result in an increase in reactive oxygen species through the increase generation of arachidonic acid. Arachidonic acid has also been shown to be an important target of reactive oxygen species. Both organic and inorganic mercury have been shown to alter calcium homeostasis but through different mechanisms. Organic mercury compounds (MeHg) are believed to increase intracellular calcium by accelerating the influx of calcium from the extracellular medium and mobilizing intracellular stores, while inorganic mercury ($Hg^{2+}$) compounds increase intracellular calcium stores only through the influx of calcium from the extracellular medium. Mercury compounds have also been shown to induce increased levels of MDA in both the livers, kidneys, lungs and testes of rats treated with $HgCl_2$. This increase in concentration was shown to correlate with the severity of hepatotoxicity and nephrotoxicity. $HgCl_2$-induced lipid peroxidation was shown to be significantly reduced by antioxidant pretreatment with selenium. Selenium has been shown to achieve this protective effect through direct binding to mercury or serving as a cofactor for glutathione peroxidase and facilitating its ability to scavenge ROS. Vitamin E has also been reported to protect against $HgCl_2$-induced lipid peroxidation in the liver.

Metal-induced carcinogenicity has been a research subject of great public health interest. Generally, carcinogenesis is considered to have three stages including initiation, promotion, and progression and metastasis. Although mutations of DNA, which can activate oncogenesis or inhibit tumor suppression, were traditionally thought to be crucial factors for the initiation of carcinogenesis, recent studies have demonstrated that other molecular events such as transcription activation, signal transduction, oncogene amplification, and recombination, also constitute significant contributing factors. Studies have shown that mercury and other toxic metals effect cellular organelles and adversely affect their biologic functions. Accumulating evidence also suggests that ROS play a major role in the mediation of metal-induced cellular responses and carcinogenesis.

The connection between mercury exposure and carcinogenesis is very controversial. While some studies have confirmed its genotoxic potential, others have not shown an association between mercury exposure and genotoxic damage. In studies implicating mercury as a genotoxic agent, oxidative stress has been described has the molecular mechanism of toxicity. Hence, mercury has been shown to induce the formation of ROS known to cause DNA damage in cells, a process which can lead to the initiation of carcinogenic processes. The direct action of these free radicals on nucleic acids may generate genetic mutations. Although mercury-containing compounds are not mutagenic in bacterial assays, inorganic mercury has been shown to induce mutational events in eukaryotic cell lines with doses as low as 0.5 μM. These free radicals may also induce conformational changes in proteins that are responsible for DNA repair, mitotic spindle, and chromosomal segregation. To combat these effects, cells have antioxidant mechanisms that work to correct and avoid the formation of ROS (free radicals) in excess. These antioxidant mechanisms involve low molecular weight compounds such as vitamins C and E, melatonin, glutathione, superoxide dismutase, catalase, glutathione peroxidase and glutathione reductase that protect the cells by chelating mercury and reducing its oxidative stress potential.

Glutathione levels in human populations exposed to methylmercury intoxication by eating contaminated fish have been shown to be higher than normal. These studies were also able to confirm a direct and positive correlation between mercury and glutathione levels in blood. They also confirmed an increased mitotic index and polyploidal aberrations associated with mercury exposure. Epidemiological studies have demonstrated that enzymatic activity was altered in populations exposed to mercury; producing genotoxic alterations, and suggesting that both chronic and relatively low level mercury exposures may inhibit enzyme activity and induce oxidative stress in the cells. There is no doubt that the connection between mercury exposure and carcinogenesis is very controversial. However, *in-vitro* studies suggest that the susceptibility to DNA damage exists as a result of cellular exposure to mercury. These studies also indicate that mercury-induced toxicity and carcinogenicity may be cell-, organ- and/or species- specific.

## Polystyrene as Hazardous Household Waste

Polystyrene (PS) is a petroleum-based plastic made from styrene (vinyl benzene) monomer. Since it was first commercially produced in 1930, it has been used for a wide range of commercial, packaging and building purposes. In 2012, approximately 32.7 million tonnes of styrene were produced globally, and polystyrene is now a ubiquitous household item worldwide. In 1986, the US Environmental Protection Agency (EPA) announced that the polystyrene manufacturing process was the fifth largest source of hazardous waste. Styrene has been linked to adverse health effects in humans, and in 2014, it was listed as a possible carcinogen. Yet, despite mounting evidence and public concern regarding the toxicity of styrene, the product of the polymerisation of styrene, PS, is not considered hazardous.

PS is a petroleum-based plastic made from styrene (vinyl benzene) monomer. Since PS was first commercially produced in 1931, it has been used for a wide range of commercial, packaging and building purposes, and it has grown to be one of the world's most ubiquitous household items. Most PS is used to make rigid durable products, such as television and computer cabinets and appliances, and nearly all of the rigid PS packaging manufactured in New Zealand (NZ) is used for food contact purposes. PS used for food packaging includes general purpose PS (GPPS) such as disposable cutlery and plates; high-impact PS (HIPS) such as yoghurt containers and single-use cold drink cups; and expanded PS (EPS) foam used as meat trays, coolers and cups.

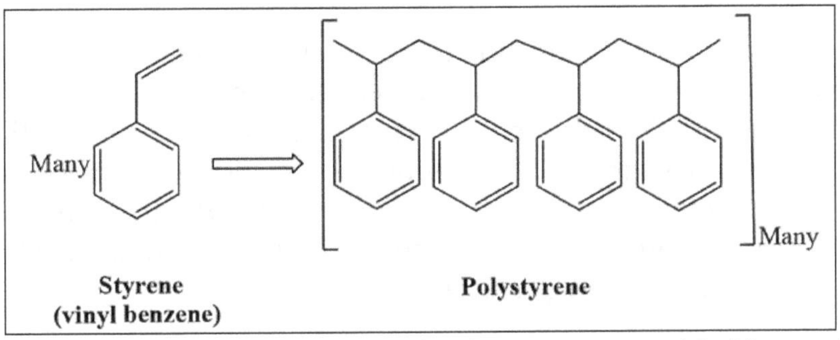


The manufacture of PS by chemically bonding many styrene (vinyl benzene) monomer units to form a long styrene polymer chain.

EPS is made of pre-expanded closed-cell foam beads. The manufacturing process involves carrying out the styrene polymerisation in droplets suspended in water. This leads to the formation of PS beads. EPS is useful because it is an excellent insulator (e.g. used to line cool boxes or chilly bins), and it absorbs shock and so it is a good moulded or bead-based packing material for transporting fragile cargo. It is used for a wide range of food contact packaging, such as meat trays, egg cartons and 'clamshell' fast food containers. EPS is often referred to by its trademark 'Styrofoam' invented by Dow Chemical in 1941. The trademark is informally used (not only in USA and Canada but also in NZ) for all foamed PS products, although strictly it should only refer to the 'extruded closed-cell' PS foam (XPS) made by Dow Chemicals and commonly used for building insulation.

While PS is popular because of its light weight and insulating properties, there is a downside to its high rate of production, consumption and disposal. In 2012, approximately 32.7 million tonnes of PS were produced globally. This figure is concerning, considering the lack of waste management of this man-made material. For example, Plastic NZ estimates that over 6784 tonnes of NZ-produced PS were consumed for packaging in NZ in 2003, with as little as 450 tonnes collected for recycling. There are currently no PS residential collection services in NZ.

In 2013, Chelsea Rochman, a scientist who studies the migration of chemicals from plastics when ingested by animals, argued for the reclassification of some key plastics as hazardous so that they could be regulated by environmental protection agencies. These key plastics are PVC, polyurethane, polycarbonate and PS.

Today, tens of towns and cities around the world prohibit the sale, possession and distribution of EPS, including Portland (Oregon, USA), Toronto (Canada), Muntinlupa (Philippines), Paris (France), and Tainan (Taiwan). While it still has been unable to enact a statewide ban, in 2016, the state of California has 65 ordinances (i.e. laws or regulations made by a local government body) on EPS. The reasons cited for ordinances banning or regulating EPS are primarily based on its potential as an environmental hazard: more specifically, a concern for the environmentally polluting potential of the physical properties of one type of PS rather than the chemical instability and potential for PS (as a family of plastics) to leak toxins into bodies and ecosystems.

Plastics harm in two key ways: chemically, when monomers, plasticisers and other hazardous PS additives leach from PS objects when poorly manufactured; physically, when PS breaks down into microplastics in the environment. The manufacture, composition and interrelationships of PS *in situ* are profoundly important in determining whether it is a 'hazard'.

Where the manufacture of PS is complete, the resulting styrene polymer has covalent (i.e. strong) bonds between the styrene monomer units; these bonds are difficult to break to release styrene. For this reason, styrene is not released from PS during environmental degradation. However, if the PS manufacturing process does not result in complete polymerisation, free styrene might remain in the PS product. This free styrene is not strongly bonded to the PS structure and so can, and likely will, leach out into the environment or into food stored in PS containers. In addition to the ability of styrene to leach from PS poorly manufactured products.

## Styrene Monomer

While the level and nature of the hazard posed by styrene have historically been hotly disputed, a growing body of evidence indicates cause for policy action to protect populations from its misuse. For example, in 2014, the International Agency for Research on Cancer (IARC) determined that styrene is a possible human carcinogen. This is based on studies in animals and emerges from research into styrene's metabolite (i.e. styrene oxide) as a chemically highly reactive epoxide, which might chemically bond to DNA and thus initiate carcinogenesis. It is now largely accepted that styrene oxide is likely to be responsible for styrene's carcinogenicity, and since there are likely to be species and inter-individual differences in metabolism, there are also likely to be differences in susceptibility to carcinogenicity between species and individuals.

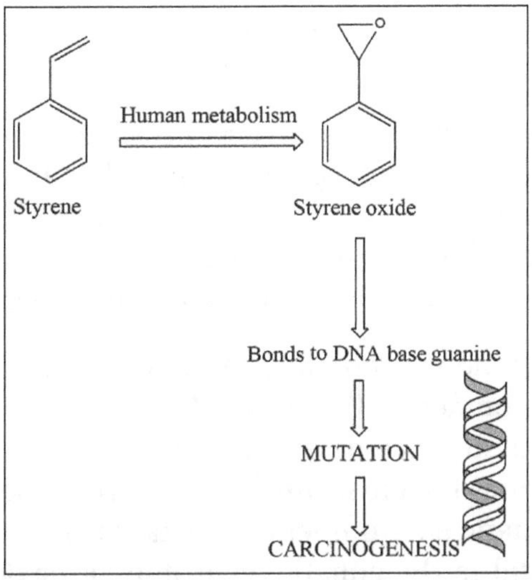


Styrene is metabolised to a highly reactive and toxic epoxide, styrene oxide, which can interact with DNA causing a mutation which might initiate carcinogenesis.

The EPA National Human Adipose Tissue Survey for 1986 identified styrene residues in 100% of samples of human fat tissue taken in 1982 in the USA, which indicates widespread exposure. In laboratory studies (e.g. in rats), styrene monomer and some other ingredients of PS have been shown to be carcinogenic and, in some cases, affect organisms in a similar way to the hormone oestrogen because of their molecular mimicry of the female hormone 17β-oestradiol and occupancy and activation of oestrogen receptors.

The National Toxicology Programme's Report of Carcinogens states that styrene is reasonably anticipated to be a human carcinogen. First listed in the Twelfth Report on Carcinogens (2011) HC CH 2 Carcinogenicity Styrene is reasonably anticipated to be a human carcinogen based on limited evidence of carcinogenicity from studies in humans, sufficient evidence of carcinogenicity from studies in experimental animals, and supporting data on mechanisms of carcinogenesis.

## The National Regional Council Concurred in 2014

Concurs with the NTP determination that there is limited but credible evidence that exposure to styrene in some occupational settings is associated with an increase in the frequency of lymphohematopoietic cancers. Additionally, the NRC report authoring committee independently reviewed the scientific evidence from studies in humans, experimental animals, and other studies relevant to the mechanisms of carcinogenesis and made level-of-evidence conclusions. Based on credible but limited evidence of carcinogenicity in traditional epidemiologic studies, on sufficient evidence of carcinogenicity in animals, and on convincing evidence that styrene is genotoxic in exposed humans, this report finds that compelling evidence exists to support a listing of styrene as, at a minimum, reasonably anticipated to be a human carcinogen.

More recently, on 22 April 2016, the Office of Environmental Health Hazard Assessment OEHHA added styrene to the list of chemicals known to the state to cause cancer. The NZ EPA (and similarly, the US EPA) determines that styrene is 'possibly carcinogenic to humans' and classifies the chemical as otherwise hazardous due to its following characteristics: flammable, acutely toxic, suspected human mutagen, carcinogen, and human reproductive or developmental toxicants, toxic to organs/systems, very ecotoxic in the aquatic environment to algal and crustaceans, slightly harmful to fish and ecotoxic to terrestrial vertebrates.

The categorisation of styrene as 'toxic' is unequivocal because, due to its high chemical reactivity, it interacts with cell systems causing widespread metabolic damage. In particular, styrene can chemically react with specific components of DNA resulting in changes, which affect the rate of division of cells. This mutation is the basis of chemical carcinogenesis and explains why styrene is a carcinogen in animal studies and reasonably anticipated to be a carcinogen in humans.

The IARC's statement that styrene is a possible human carcinogen is reinforced in statements made by the USA and NZ Environmental Protection Agencies (EPAs). The USA and the NZ EPA, among other EPAs globally, classify styrene as 'hazardous' under a wide variety of ecotoxic and toxic categories. The NZ EPA rates the environmental hazards associated with styrene as follows: (algal): very ecotoxic in the aquatic environment; (crustacean): very ecotoxic in the aquatic environment; (fish): slightly harmful in the aquatic environment or otherwise designed for biocidal action; ecotoxic to terrestrial vertebrates. In addition, the European Union currently places styrene in Category 1 in their list of potential endocrine disruptors. These new categorisations align with a growing global interest in the hazardous nature of the ingredients used in the production of plastics. Interestingly, according to a hazard-ranking model based on the United Nations' Globally Harmonized System of Classification and Labelling of Chemicals, the chemical ingredients of more than 50% of plastics have now been determined to be hazardous.

## The Polymerisation of Styrene

Despite building evidence and public concern regarding the toxicity of styrene, the product of the polymerisation of styrene, PS, is not listed as 'hazardous' in any policy documents. Regardless, some community groups and a growing number of independent scientists now treat PS as a hazardous waste item. When styrene is fully polymerised in the manufacturing process to form PS, the reactivity of the styrene component of the polymer is removed completely because the chemical bond that forms between the monomer units changes the nature of the reactive moieties of the styrene molecule. This explains why PS has very low mammalian toxicity. In addition, the monomer units are very strongly bonded together which means that styrene monomer cannot be released, in an environmental context, from its polymerised form. The bacterial degradation (e.g. in landfill) of PS does not liberate styrene, but rather produces substances such as 4-phenylvaleric acid which is of low toxicity.

However, if the styrene polymerisation process used in the manufacture of PS is *not* complete, styrene monomer might contaminate the PS (termed 'residual monomer'). This styrene can migrate into food packed in styrene-contaminated PS. Since styrene is reasonably fat soluble ($LogP_{ow}$ [styrene] = 3.6; i.e. styrene is $10^{3.6}$ times more soluble in fat than water), migration is greater in fatty foods (e.g. milk) in PS containers. On these solubility grounds, migration of styrene into a water-based ecosystem is less likely than into fat-containing foods, but taking account of the vast quantities of water in the environment, even at low water solubility, significant transfer of styrene to aqueous ecosystems will occur. Once in the aquatic environment, styrene will be rapidly absorbed via the lipid-based cell membranes of aquatic organisms and will concentrate up the food chain. Thus, styrene is more likely to affect animals at high trophic levels (i.e. predators).

Other chemicals (e.g. plasticisers) are sometimes added to PS to modify its physical properties for particular applications. For example, tris(4-nonyl-phenyl) phosphite is sometimes added as an antioxidant to prevent PS degradation. Such molecules can leach from PS and contaminate both the environment and products (e.g. food) stored in PS containers.

Regardless of whether styrene is completely polymerised or not during the manufacture of PS, all PS waste have significant implications for the environment. This is because the complete environmental degradation of PS is very slow and produces small PS particles *en route*. These PS particles have significant implications in ecosystems as they build up in, for example, marine environments. In addition to the physical hazards they pose, hydrophobic plastics such as PS are the most hazardous of plastics in fresh water and marine ecosystems because of their ability to adsorb persistent organic pollutants (POPs)—such adsorbed POPs can be released following ingestion of PS microparticles by animals (e.g. fish).

## Polystyrene as an Environmental Hazard

While PS is highly valued for its light weight, strength, thermal insulation and shock absorbing properties, its production and disposal reveal significant threats to the environment. Today, most plastic waste goes to landfills where chemicals can leach from the plastic and contaminate soil and groundwater. The US EPA reports that 'each year Americans throw away 25,000,000,000 [25 billion or 47,565 per minute] Styrofoam cups. Even 500 years from now, the foam coffee cup you used this morning is likely to survive intact in a landfill'.

There is no domestic kerbside collection of any form of PS provided by councils in NZ. The lack of PS recycling services available to householders in NZ and the volume of EPS make disposal via council rubbish bags expensive. Consequently, littering or 'fly tipping' and household Burning of PS can become residential alternatives, thus raising the likelihood of releasing PS and its potentially toxic combustion products into the environment.

## Ecotoxicity in the Manufacture of Polystyrene

In 1986, a US EPA report on solid waste declared PS manufacturing the fifth largest source of hazardous waste in the USA. In addition, the manufacture of PS is energy intensive, creating large amounts of greenhouse gases (e.g. $CO_2$) and liquid and solid waste. To add to this life-cycle-based harm to the environment, PS is manufactured from petroleum: a non-sustainable and heavily polluting resource. Consequently, the environmental production costs of PS have been ranked the second worst in the USA by the California Integrated Waste Management Board.

It is possible that differences in manufacturing practices might lead to different levels of monomer residues remaining in the final product. For example, very preliminary studies on polycarbonate plastics showed that very different amounts of bisphenol A (BPA) monomer leached from different plastic products manufactured in Korea, China and NZ. While very preliminary, this might reflect differences in manufacturing processes and controls between these countries. As Asia becomes the region of choice (on economic grounds) for manufacture of many plastic products, it is possible that PS products made with residual plastic monomers will increase. Consequently, the risk of monomer leaching into the environment and into products stored in plastic containers will rise as a result. The increase in free trade of these PS objects across geopolitical boundaries means that regulating the manufacture and responsible disposal of PS is very complex indeed.

## Physical and Ecotoxic Threats in Marine Environments

Plastic pollution in marine environments has become so dire that the term 'plastic pollution' is now synonymous with 'marine pollution'. A 2016 World Economic Forum report predicts that the ratio of plastic to fish in the ocean is expected to be 1:3 by 2025. Currently, 60–80% of waste found in marine environments is plastic, and in 2014, it was estimated that more than 226,796 tonne of plastic is currently afloat at sea. Because of its light weight, PS is highly mobile and can transport invasive species across marine boundaries. PS is prevalent in NZ coastal areas: for example, in a 2016 study of microplastics on Canterbury's (NZ) coastlines, the majority of plastics (55%) found were PS.

Douglas McCauley, a marine biology professor at the University of California, Santa Barbara, USA studied the mechanical and chemical causes of harm to marine animals from EPS. The environmental degradation of PS (e.g. by bacteria) leads to the production of small fragments (microplastics) which survive for a very long time in the environment (e.g. in marine systems) and cause physical effects. For example, fish mistake the PS particles for food and eat them. This leads to malnutrition as PS provides fish with no nutritional benefit while making them feel full, thus suppressing a desire to eat. Plasticisers and other additives are less common in PS products than in other plastics (e.g. polyvinyl chloride [PVC]). However, the antioxidant tris(4-nonylphenyl) phosphite sometimes used in PS products is a potent oestrogen mimic and is part of a cocktail of estrogenic environmental contaminants that is thought to be responsible for male feminisation in animals and humans. To be clear, styrene is *not* produced by the environmental degradation of PS but by the leaching of residual styrene from incomplete polymerisation.

As outlined above, there is a further, very important, property of PS that has a significant bearing on its environmental and human toxicity: its extreme hydrophobicity. Since, in a chemical context, like chemical properties attract like, PS attracts and adsorbs (i.e. sequesters on its surface) other hydrophobic molecules (e.g. POPs). This is important because it means that micro PS particles (produced as part of the environmental degradation of PS) will sequester and transport POPs and other hydrophobic toxins in the aquatic environment. If PS microparticles are ingested by animals (or humans), the sequestered POPs might be stripped from the PS and absorbed into the animal's system. This makes PS an excellent vector for highly toxic hydrophobic chemicals.

Seabirds that have consumed plastic waste have been found to have polychlorinated biphenyls (PCBs) and POPs in their tissues at 300% greater concentrations than in similar birds that have not eaten plastic. PS is particularly good at attracting oily (i.e. hydrophobic) chemicals such as PCBs, flame retardants such as polybrominated diphenylethers (PBDEs), pesticides (e.g. DDT) and surfactants (e.g. 4-nonylphenol an endocrine disruptor). These chemicals have been estimated to be adsorbed by PS at concentrations up to a million times greater than in the surrounding water.

The chemicals PS attracts are regarded as 'priority pollutants':

Chemicals that are regulated by government agencies, including the US EPA, because of their toxicity or persistence in organisms and food webs. These chemicals can disrupt key physiological processes, such as cell division and immunity, causing disease or reducing organisms' ability to escape from predators or reproduce.

Rochman and her team found that at least 78% of priority pollutants listed by the EPA and 61% listed by the European Union (EU) were associated with plastic debris (either ingredients of plastic or adsorbed from the environment). PS microplastics contaminated by these POPs enter the food chain when eaten by marine species and might end up on our dinner tables at home.

Taking all of these chemical and toxicological properties of PS into account, PS *per se* is one of the less problematic plastics in a purely toxicological sense. However, its high rates of production, poor waste management, slow environmental degradation to form microparticles and adsorption of hydrophobic toxic chemicals have led, at the macro level, to huge quantities of PS waste presenting physical problems at all environmental levels, and, at a micro level, PS transporting adsorbed toxins to unsuspecting marine ecosystem consumers. It is clear that a policy connection has not yet been made between these hydrophobic polymer 'carriers' and priority pollutants in the marine environment. If there were, it would be more likely that there would be cause to categorise hydrophobic plastics as 'hazardous.' Growing evidence, however, continues to highlight the previously unforeseen and complex relational behaviours of PS and its additives with other materials and bodies in marine environments. This growing evidence supports Liboiron's argument that the governance of plastic pollution via 'safe levels' may not adequately capture the 'afterlives' of plastic polymers.

## Polystyrene Combustion

The incomplete combustion (at temperatures that equate to household burning) of PS produces myriad products including styrene, PAHs, including fluoranthene which has been shown to be carcinogenic in mouse studies and the IARC, have classified it as a Group 3 carcinogen (i.e. not classifiable as to its carcinogenicity to humans). Concerns have also been raised about PS as a hazard during house or commercial building fires or the disposal of PS by burning due to the toxins it releases when PS is combusted, particularly when incinerated as household waste in residential areas. The National Bureau of Standards Center for Fire Research identified 57 chemicals released during the combustion of EPS. Of these, perhaps the most toxic include, PAHs, carbon black (i.e. the copious soot produced when PS burns) and carbon monoxide.

The use of extruded XPS and EPS in building construction is also a focus of concern due to the persistence, toxicity and ecotoxicity of the brominated flame retardant HBCD used in the manufacture of these PS foams. In response to these growing concerns, the EU (under the Stockholm Convention) banned the use, import or export of HBCD on 26 November 2015.

## Polystyrene and Food Safety

PS can be hazardous to human health in the following circumstances: eating marine animals contaminated with the residual monomers and additives in PS and the POPs adsorbed onto PS; inhalation of gases created when PS is heated or combusted; chemical exposure in the manufacture of PS; and exposure to tropospheric ozone caused by HFCs in the manufacture of PS.

New scientific research is raising some doubt about the safety of food when in contact with PS under certain conditions. One of those conditions has been recognised in the US FDA guidelines in contact with fat above ambient temperature. In 2014, the European Food Safety Authority (EFSA)'s Panel on Food Contact Materials, Enzymes, Flavourings and Processing Aids (CEF) conducted a study requested by the Food Standards Agency (FDA), UK. They concluded the following: substances (butadiene, ethyl acrylate, methyl methacrylate, styrene) copolymer, (butadiene, ethyl acrylate, methyl methacrylate, styrene) copolymer crosslinked with divinylbenzene and (butadiene, ethyl acrylate, methyl methacrylate, styrene) copolymer crosslinked with 1,3-butanediol dimethacrylate, in nanoform do not raise a safety concern for the consumer if used as additives individually or in combination at up to a total of 10% w/w in non-plasticised PVC used in contact with all food types at ambient temperature or below including long-term storage.

However, household PS is not always used in contact with food at ambient temperatures; food is often heated in PS containers in microwave ovens; polystyrene cups and food containers are most commonly used for hot beverage consumption. This again raises the question of safe levels of styrene monomer and safe conditions of use when PS items are exposed to a wide range of untested conditions introducing a toxic risk.

Researchers have reported that volatile styrene monomer was found in shells of eggs after they were stored for 2 weeks in PS containers at supermarkets. Dishes cooked with these contaminated eggs contained seven times more ethylbenzene and styrene compared to those prepared from fresh farm eggs not packaged in PS. Not surprisingly, the American Chemistry Council (representing the plastics industry) vehemently denies any health risks posed by PS food packaging to the US public.

## References


- Introductory-chapter-introduction-to-hazardous-waste-management, management-of-hazardous-wastes: intechopen.com, Retrieved 11 March, 2019
- The-4-types-of-universal-waste: heritage-enviro.com, Retrieved 10 July, 2019
- Biodegradable-and-non-biodegradable, chemistry: byjus.com, Retrieved 26 May, 2019
- Time-classify-plastics-hazardous-wastes: redbags.com, Retrieved 08 February, 2019
- Environmental-effects-plastic-pollution, environment: eartheclipse.com, Retrieved 17 May, 2019
- Polystyrene-as-hazardous-household-waste,household-hazardous-waste-management: intechopen.com, Retrieved 23 August, 2019

# 2

# Understanding Hazardous Waste Management

A system which is used for the management, collection, treatment and disposal of hazardous waste to protect the environment is called a hazardous waste management system. Radioactive waste management refers to treating, storing, and disposing liquid, airborne, and solid effluents from the nuclear industry's operations. This chapter has been carefully written to provide an easy understanding of the varied facts of hazardous waste management.

Hazardous-waste management is the collection, treatment, and disposal of waste material that, when improperly handled, can cause substantial harm to human health and safety or to the environment. Hazardous wastes can take the form of solids, liquids, sludges, or contained gases, and they are generated primarily by chemical production, manufacturing, and other industrial activities. They may cause damage during inadequate storage, transportation, treatment, or disposal operations. Improper hazardous-waste storage or disposal frequently contaminates surface and groundwater supplies. People living in homes built near old and abandoned waste disposal sites may be in a particularly vulnerable position. In an effort to remedy existing problems and to prevent future harm from hazardous wastes, governments closely regulate the practice of hazardous-waste management.

It involves reducing the amount of hazardous substances produced, treating hazardous wastes to reduce their toxicity, and applying sound engineering controls to reduce or eliminate exposures to these wastes. From a chemical engineering perspective, any wastes, but especially hazardous wastes, are produced unintentionally. Facilities that generate the chemicals are attempting to produce substances demanded by the marketplace, but in the process and under the wrong circumstances these substances or their components become hazards. If the same substance were produced intentionally, it would be a hazardous material, but not a hazardous waste.

Engineers design and operate facilities to prevent, contain, decontaminate, and otherwise treat these wastes before they are released. However, hazardous waste engineering is most often associated with interventions after the wastes have been released. Engineers are frequently called upon to address these wastes once they are released into the environment.

A hazard is expressed as the potential of an unacceptable outcome. For chemicals, the most important hazard is the potential for disease or death. Epidemiologists refer to these as morbidity and mortality, respectively. The hazards to human health are referred to collectively in the medical and environmental sciences as toxicity. Human toxicology is the study of these health outcomes and their potential causes. Likewise, the study of hazards to ecosystems is known as ecological toxicology or simply ecotoxicology, which is further subdivided by discipline, for example, aquatic toxicology and mammalian toxicology.

## Evaluating Hazards

In the United States, as in many countries, the term hazardous waste is a legal term defined by the regulatory agencies. Under the RCRA hazardous waste regulations, the U.S. Environmental Protection Agency (EPA) is principally responsible for the permitting of hazardous waste treatment, storage, and disposal facilities. The EPA, however, has authorized most states to operate portions or all of their hazardous waste programs. This differs from Subtitle D, nonhazardous wastes. In fact, the U.S. Congress intended that permitting and monitoring of municipal and nonhazardous waste landfills be a state responsibility.

The EPA has developed standard approaches and set criteria to determine if substances exhibit hazardous characteristics. Because RCRA defines a hazardous waste as a waste that presents a threat to human health and the environment when it is improperly managed, the government identified a set of assumptions that would allow for a waste to be disposed if it is not subject to the controls mandated by Subtitle C of RCRA. This mismanagement scenario was designed to simulate a plausible worst case.

Under a worst-case scenario, a potentially hazardous waste is assumed to be disposed along with municipal solid waste in a landfill with actively decomposing substances overlying an aquifer. Hopefully, few wastes would be managed in this manner but a dependable set of assumptions are needed to ensure that the hazardous waste definition was implemented. The Hazardous and Solid Waste Amendments of 1984 (HSWA) established the Toxicity Characteristic Leaching Procedure (TCLP) to provide replicable results for contaminants commonly found in hazardous waste sites. The specific medium or substrate used in the test is dictated by the alkalinity of the waste. The liquid extracted from the waste is analyzed for the 39 listed toxic constituents and the concentration of each contaminant is compared to the TCLP standards specific to each contaminant.

Table: Toxicity Characteristic Chemical Constituent Regulatory Levels for 39 Hazardous Chemicals.

| Contaminant | Regulatory Level (mg $L^{-1}$) |
|---|---|
| Arsenic | 5.0 |
| Barium | 100.0 |
| Cadmium | 1.0 |
| Chromium | 5.0 |
| Lead | 5.0 |
| Mercury | 0.2 |
| Selenium | 1.0 |
| Silver | 5.0 |
| Endrin | 0.02 |
| Lindane | 0.4 |
| Methoxychlor | 10.0 |
| Toxaphene | 0.5 |
| 2,4-D | 10.0 |
| 2,4,5 TP (Silvex) | 1.0 |
| Benzene | 0.5 |
| Carbon tetrachloride | 0.5 |

| Chlordane | 0.03 |
|---|---|
| Chlorobenzene | 100.0 |
| Chloroform | 6.0 |
| *o*-Cresol | 200.0 |
| *m*-Cresol | 200.0 |
| *p*-Cresol | 200.0 |
| Cresol | 200.0 |
| 1,4-Dichlorobenzene | 7.5 |
| 1,2-Dichloroethane | 0.5 |
| 1,1-Dichloroethylene | 0.7 |
| 2,4-Dinitrotoluene | 0.13 |
| Heptachlor (and its hydroxide) | 0.008 |
| Hexachloroethane | 3.0 |
| Hexachlorobutadiene | 0.5 |
| Hexachloroethane | 3.0 |
| Methyl ethyl ketone | 200.0 |
| Nitrobenzene | 2.0 |
| Pentachlorophenol | 100.0 |
| Pyridine | 5.0 |
| Tetrachloroethylene | 0.7 |
| Trichloroethylene | 0.5 |
| 2,4,5-Trichlorophenol | 400.0 |
| 2,4,6-Trichlorophenol | 2.0 |
| Vinyl chloride | 0.2 |

Hazards other than toxicity are also important and regulated accordingly. These include threats to public and personal safety, which may result from a substance's potential to ignite, its corrosiveness, flammability, or explosiveness. A substance may be hazardous to public welfare if it damages property values or physical materials; for example, expressed as its corrosiveness or acidity. The hazard may also relate to environmental quality, such as an ecosystem stress indicated by diminished productivity and biodiversity, loss of important habitats, and decreases in the size of the population of sensitive species. The hazard may be inherent to the substance. But more than likely, the hazard depends upon the situation and conditions where the exposure may occur. The substance is most hazardous when a number of conditions exist simultaneously, such as the hazard to firefighters using water in the presence of oxidizers.

Since 1980, specific methods for testing of hazards are delineated in the EPA publication SW-846, Test Methods for Evaluating Solid Waste, Physical/Chemical Methods, which is a compendium of analytical and sampling methods that comply with the RCRA regulations. The SW-846 provides guidance, although not necessarily a legal requirement, when conducting RCRA-related sampling and analysis requirements. This guidance is updated and revised as new information become available.

## Waste Management Hierarchy

A typical sequence for managing hazardous wastes involves source reduction, recycling, treatment, and disposal. The waste generator is encouraged to begin at the top of the hierarchy and proceed downward, if necessary (obviously, if the waste is eliminated, there is no further need to follow the sequence). In the United States, the term pollution prevention is often preferred to source reduction, and waste minimization is used as an umbrella term for strategies and technologies that use source reduction and recycling techniques.

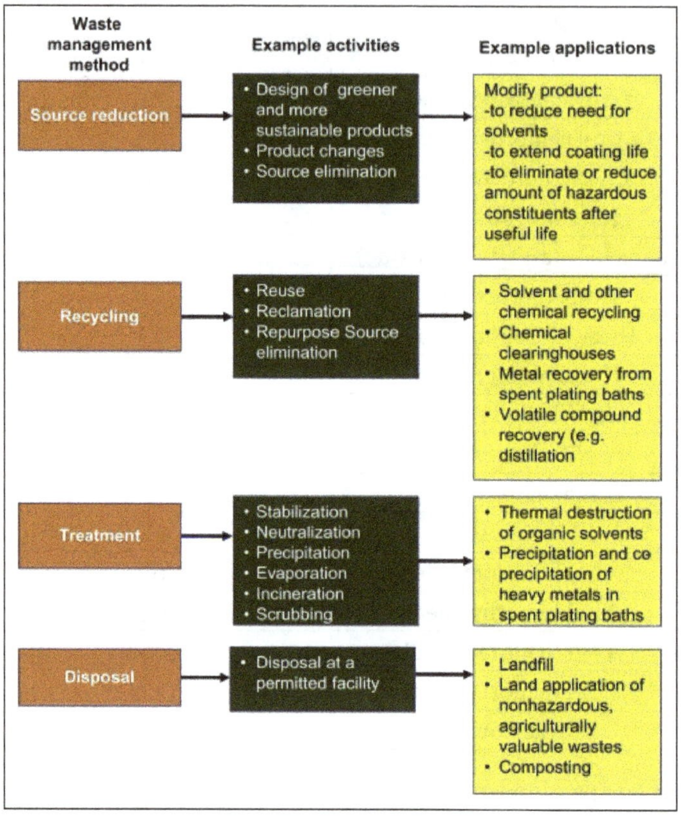


Flowchart for management of waste: reduction from manufacture, use, treatment, and disposal.

## Source Reduction

Reducing the amount or severity of a hazardous can be achieved via process modifications, feedstock substitutions, improvements in feedstock purity, changes in housekeeping and management practice, increases in the efficiency of equipment, and recycling within a process. In the United States, there is a requirement in the federal environmental regulations that all large hazardous waste generators must have a program at their facility to encourage source reduction and recycling. Other countries, such as Austria, Germany, and Denmark, have initiated direct subsidies to encourage preferable waste management options.

## Recycling

The use or reuse of hazardous waste as an effective substitute for a commercial product or as an ingredient or feedstock in an industrial process can not only be advantageous from a public health and environmental standpoint, but also from a financial perspective. Recycling includes the reclamation of useful constituent fractions within a waste material or the removal of contaminants from a waste to allow it to be reused.

## Disposal and Treatment

Hazardous waste disposal consists of the discharge, deposit, injection, dumping, spilling, leaking, or placing of hazardous waste into or on any land or body of water so that the waste or any constituents may enter the air or be discharged into any waters, including groundwater.

The portion of the hazardous waste that cannot be eliminated or recycled must undergo a method, technique, or process that changes the substance's physical, chemical, or biological character to neutralize it; to recover energy or material resources from the waste; or to render the waste nonhazardous, less hazardous, safer to manage, amenable for recovery, amenable for storage, or reduced in volume.

## Hazard Incident Response

For example, in the United States, the Occupational Health and Safety Administration (OSHA) is interested in protecting workers from exposure to hazards at waste sites. This calls for an incident response model that consists of five components:

- Recognition,
- Evaluation,
- Information,
- Safety,
- Control.

Before entering a potentially dangerous situation at a waste site, one needs to recognize potential hazards, including the physical, chemical, and biological hazards. However, this is not enough for safe entry. An evaluation of the site depends on sound information. This can be drawn from current and previous site investigations; interviews with suppliers, customers, employees; and scientifically sound sampling and monitoring before, during, and after any incident response. Controls include containment, suppression, and treatment and depending on the severity of a hazard, for example, immediate evacuation from a potentially explosive situation, or shutoff of a contaminated drinking water supply.

This process may involve either physical or health hazards. Physical hazards include reactivity, fire, explosion, corrosivity, and radioactivity. Health hazards include various types of effects, including acute effects like poisoning to chronic effects such as cancer or neurotoxicity.

Anyone involved in any emergency or incident response must be properly trained to recognize these hazards. Thus the EPA and OSHA require completion of an intensive course in hazardous waste operations and emergency response. Employers must provide this training and must ensure that workers be recertified annually.

## Management of Hazardous Waste in Developing Countries

In developing countries, hazardous waste management systems lack a systematic approach to administer waste management programmes; inability to effectively collect and manage wastes as well as to reduce the negative impacts of those activities. The current regulatory frameworks and regulations do not adequately address hazardous waste treatment and final disposal. There are inadequacies in the implementation of regulations associated with hazardous waste management due to fragmented responsibilities among government departments and local authorities.

The treatment of hazardous waste should take place under regulated and controlled conditions. Hazardous waste management includes the possession, transportation, handling, storage, and ultimate disposal of waste. However, in developing countries the treatment of hazardous waste takes place in unregulated or uncontrolled conditions, and in some cases hazardous waste are exported to developing countries by the developed countries. It has been established that for most parts, hazardous wastes are treated in unlicensed facilities using conventional methods such as landfilling. The remainder of the waste stream is treated at authorized facilities with low technologies and low environmental standards. These include incineration plants, landfills and oil recovery. An important fraction of hazardous wastes is still mixed with non-hazardous wastes, being mainly landfilled and producing serious environmental impacts regarding heavy metals and persistent organic pollutants content in landfill leachate, thus creating a great health risk to municipal workers, the public, and the environment. The main problems affecting the management of hazardous waste in developing countries include the following:

- Lack of necessary rules, plans, regulations, and instructions on different aspect of collections and disposal of waste.
- Lack of policy directions or incentives for existing local authority or private sector landfill operators.
- Inadequate Institutional capacities.
- Lack of hazardous waste prevention activities.
- Shipment of hazardous to developing countries from developed countries.
- Mixing of hazardous waste with domestic waste or commercial waste.
- Lack of collection facilities for hazardous waste.
- Failure to quantify the hazardous waste generated in reliable records.
- Existing environmental permits mostly fail to contribute to Best Available Technology (BAT).
- Not all regional environmental authorities register the generated waste types in the permit using waste classification codes.
- Information about waste production rates corresponding to the recorded waste generation is usually missing. The developing countries waste information system is not detailed enough to gather information about concentrations of hazardous substances, only total mass.
- Lack of enforcement of and compliance with the existing regulations.

- Lack of priority sectors for prevention of hazardous waste.
- Lack of conformance with the best international practices on hazardous waste management.
- Inadequate infrastructure and self-sufficiency in hazardous waste management.
- Absence of dedicated waste managers and committees, as well as plans responsible for monitoring hazardous waste practices.
- Absence of dedicated national facilities.
- Lack of guidance, awareness, capacity building training on the management of sectoral hazardous waste management.
- Absence of specific policies dedicated to the management of hazardous household waste.

This suggests that integrated strategies for hazardous wastes recovery are needed in the developing country industry that may reduce the disposal rate of these wastes in communal landfills and impulse resource recovery and recycling of valuable materials of these wastes. Therefore, it is necessary to develop high technologies; promoting recovery and recycling centers and hazardous waste management strategies, which are environmentally, socially, economically, and technically feasible.

Doing so recognizes the proximity of best international practices to manage hazardous waste: maximizing the reuse, recycling, and recovery of potential materials, precious metals and where practical through the provision of a range of local treatment options; ensuring the availability of recovery and disposal outlets and stimulating green economy opportunities within nations.

Currently, many developing countries have no dedicated hazardous landfill disposal facilities. A few countries such as South Africa have hazardous landfill disposal facilities. However, in many developing countries there are considerable challenges to develop hazardous waste landfills, including social acceptance, regulatory, technically skilled manpower, financial resources, limited technology or provision of such infrastructure. It is demonstrated that while landfill is the least favoured option on the waste hierarchy, it is recognized that for some non-recoverable or non-combustible hazardous wastes it will need to be considered.

Satisfactory infrastructure and monitoring of movement of hazardous waste are non-existent. Indiscriminate disposal of solid waste materials such as e-waste, used oil, poses major environmental problems including the soil contamination and threat to animals. Hazardous wastes such as wastewater from healthcare facilities are often discharged into storm channels, which are not periodically cleaned. Monitoring of the health and environmental risks associated with these practices is not done due to the lack of technically skilled manpower, health and safety personnel.

Also, scientific, engineering, and organizational challenges, which need to be taken care, are given below What is so worrying here is that the recycling base in the developing countries is very weak.

Table: Scientific, engineering and organizational challenges in developing countries.

| Engineering challenges | Scientific challenges | Organizational challenges |
|---|---|---|
| • Scientific collection, transportation, segregation, and disposal of hazardous waste<br>• Practical techno-economical solutions<br>◦ Recycling<br>◦ Reuse<br>◦ Recovery | • Value addition of recyclables for reuse<br>• Disposal of process waste and residues<br>• Eco-friendly recovery solution | • Organization and structuring hazardous waste management system<br>• Training and awareness |

## Industrial Hazardous Waste

Industrial hazardous waste is defined as waste generated from industrial sectors and pose immediate danger to the environment and the public. Industrial hazardous wastes are characterized in terms of toxicity (acute, chronic, and extrinsic), inflammability, reactivity, and corrosiveness. The main important sources of industrial hazardous wastes are mining, chemical, mechanical, pulp and paper industries, cement production facilities, wood remanufacturing facilities, etc. Important industrial hazardous wastes include used oil and oil contaminated materials, spent solvent. Industries in developing countries such as Lebanon are estimated to generate 3000 to 15,000 tons/year of hazardous wastes due to deficient physical infrastructure and absence of environmental management plans, which will inevitably amplify environment impacts associated with industrial activities. Nevertheless, there is limited data on the quantities of industrial hazardous waste, where is going, where it is generated and disposed of. However, due to the accelerated development of economies, mass manufacturing and processing industries and less strict standards on environmental quality assessment, the quantities of industrial hazardous waste (IHW) will continuously increase. Meanwhile, in developing countries it has been confirmed that an important fraction of hazardous waste is still mixed with non-hazardous wastes, mainly through landfilling, and producing serious environmental impacts regarding heavy metals and persistent organic pollutants content in landfill leachate. Scientific research has observed that wastes are not separated in origin into hazardous and non-hazardous materials. Furthermore, non-hazardous are also not separated into recyclables, non-recyclables, and domiciliary. In fact, scientific research has found that developing countries lack regulations that specifically deal with industrial hazardous wastes. In addition, in the developing countries there are no hazardous waste management plans and authorized facilities to manage, treat, and eliminate hazardous wastes. The separation of all the waste streams for possible waste reuse is not implemented in the developing countries. This suggest that there is an urgent need to establish policies geared towards stimulating industries, commerce and societies to manage hazardous waste in a sustainable manner; encouraging the shift from traditional waste management practices to "cradle-to-cradle", and a reduction of the problem of management of hazardous wastes.

## Medical Waste

Medical waste is waste generated from health-related facilities such as hospitals, clinics, health centre, research institute, etc. It contains toxic chemicals, heavy metals, and may contain substances that are genotoxic or radioactive, pathological waste, sharps (e.g. needles, syringes, scalpels, knives, broken glass, etc.), infectious waste, bulk human blood and blood products. Presently, in developing countries, there is no statistical data on medical waste. However, in most cases, the estimated quantities of medical waste are extrapolated using bed occupancy and number of beds in health institutions. For instance, in Bangladesh, Dhaka City, it is reported that 37±5 tonnes of medical waste is generated from hospitals, clinics, and other healthcare facilities. However, scientific studies have shown that secure disposal of medical waste remains quite lower in several developing countries. Meanwhile, significant fraction of medical waste is disposed as municipal solid waste (MSW) or discharged without monitoring and control; incineration is the only predominant formal way to treat medical waste. In addition, in developing countries, majority of the incineration facilities are not fully maintained and operational; use primitive technologies and simple equipment, which may consequently cause pollution.

## Household Hazardous Waste

In the home, it has been pointed out that there are several jobs which generate hazardous waste because of the product used may contain hazardous substances. It has been confirmed that such products include paints, cleaners, stains and varnishes, car batteries, motor oil, and pesticides. As a result, the used leftover contents from such consumer products are known as household hazardous waste. Other HHW include e-waste. It is estimated that developing countries will discard 400-700 million obsolete computers by 2030. Most of the waste is sent to Africa or Asia under false pretences as donations from developed countries, especially Europe and the United States. Currently, in developing countries, it has been observed that a few national and local regulations have been developed and/or drafted for HHW management. Due to the lax enforcement and monitoring, valuable hazardous wastes such as lead batteries, waste mineral oils, photographic chemical wastes, waste mercury lamps, and certain electronic waste containing heavy metals and printed circuit boards (PCBs) are improperly disposed of, thereby causing significant negative impacts on public health and the environment. Furthermore, in developing countries some HHW is almost managed as municipal solid waste. The informal recyclers are the dominant sector recovering and recycling valuables. Most importantly, in developing countries, safe disposal methods of hazardous wastes including incineration, landfilling, and special treatment (e.g. microwave, sterilization, etc.) are not well-founded. Thus, this inevitably accelerates pollution of HHW.

Table: Hazardous wastes and country practice.

| Waste | Practice | | | | | | |
|---|---|---|---|---|---|---|---|
| | Equipment availability | Legal framework | Public awareness | Technology | Inventory | Separation of hazardous waste | Incentives or penalties |
| HHW | – | Nascent or non-existent | Lack of capacity and awareness | Low technology incineration | Lack of information on HHW | Unfriendly environmental collection, treatment and disposal system | Low incentives or penalties |
| Medical waste | – | No specific policy | Lack of capacity and awareness | Low technology incineration | Lack of information on medical waste | Inadequate collection, treatment and disposal system | Low incentives or penalties |

| Industrial hazardous waste | – | No specific policy | Lack of capacity and awareness | Low technology incineration | Lack of information on industrial waste | Comingled with MSW | Low incentives or penalties |
|---|---|---|---|---|---|---|---|

## Prevention of Hazardous Waste

Scientific research studies have proven that in waste management, prevention is at the top of the hierarchy and represents the preferred policy approach to materials management and an alternative to reduce the wastage of materials or resources. As a result, it is substantiated that prevention of waste is preferable to its generation and to the monetary and environmental costs incurred as a result of its generation. It is proven that radioactive waste will continue to be a priority in order to achieve greater resource efficiency. For this reason, an integrated approach should be adopted to design programs, and agencies to lead and coordinate a wide range of prevention initiatives to reduce the potential environmental and public health impacts. These should include development of programs focusing on a number of strategic objectives in order to achieve the goal of prevention which will include the prevention of radioactive and hazardous wastes. This suggests that to reduce the public health and environmental impacts of radioactive and hazardous wastes, a range of regulations and innovative processes must be developed to control the content of these potentially hazardous and harmful substances.

## Innovative Processes used for Treatment of Hazardous Wastes in Developing Countries

The negative environmental and public health impacts caused by hazardous wastes as well as the use of complex equipment, infrastructure, sophisticated controls, and dangerous processes have encouraged developing countries to develop innovative processes to treat these wastes. These innovative processes play a critical role in the final treatment of hazardous wastes, including the protection of the surrounding soil where these wastes are disposed of, and such technologies are cost-effective, easy to operate in the face of limited infrastructure, technical knowledge and expertise, and suitable for developing countries. One of these innovative processes is phytoremediation. Phytoremediation involves the planting of trees to prevent and repair environment from hazardous wastes, including restoration of degraded soil, and conservation. It is used in China, India, Pakistan, etc. However, scientific research has found that phytoremediation has limitation of long lasting and of low efficiency. Therefore, there is an urgent need for long lasting and high efficiency processes for the treatment of hazardous wastes.

## Best Practices of Hazardous Waste Management in the World

It is important to note that policy makers, regulators, product producers, generators, and holders of hazardous waste need to play a critical role in ensuring that such materials are prevented, minimized, collected, and treated properly in accordance with the waste hierarchy or "cradle-to-cradle". The best practices in managing radioactive and hazardous waste include the following:

- Adopting alternative strategies (e.g., physico-chemical treatment).
- Treatment with or without heat recovery. For instance, in Chile, used oil and spend solvents are used as raw materials for alternative fuels production.

- Making new regulations where necessary and appropriate.
- Ensuring that other government departments and public bodies fulfill their roles and responsibilities identified in hazardous waste management.
- Facilitating a two-way communication with sectoral and stakeholder interests.
- Municipalities fulfill their important role in providing small-scale collection services and generally raising awareness in hazardous waste management.
- Prevention: Engaging with priority sectors on hazardous waste prevention waste activities through cleaner technologies and better compliance with regulation.
- Devising future new policies on the management of hazardous waste by also taking into consideration the generation rates of these types of wastes and potential savings from recovery of these wastes.

### Recommendations for Developing Countries towards Hazardous Waste Management

The environmentally sound management of hazardous wastes is becoming a major concern in developing countries due to the diversity of the waste stream and toxic material within it, as well as the negative environmental and public health impacts caused. Hence, several practical recommendations are suggested which include the following:

- Creation of public awareness of the potential of recycling hazardous wastes.
- Source reduction.
- Capacity building and human resources development for hazardous wastes recycling.
- Monitoring and evaluation of hazardous wastes management systems as well a reporting programmes.
- Development of appropriate infrastructure, technical knowledge, and expertise.
- Strengthening and reforming existing regulatory frameworks.
- Development of decision support for identifying appropriate technologies for treatment of hazardous wastes.
- Provision of funding for identification of emerging best and state-of-the art technologies.
- Development of regional hazardous waste management system.

## Radioactive Waste Management

Radioactive waste is any material that is either radioactive itself or is contaminated by radioactivity, for which no further use is envisaged. Government policy means that certain nuclear materials such as uranium, plutonium and spent nuclear fuel have not been declared as wastes by their owners.

## Categories of Radioactive Waste

The term 'radioactive waste' covers a wide variety of material, ranging from wastes that can be put safely into a dustbin to items that need remote handling, heavy shielding and cooling to be managed safely. To be clear, radioactive waste is considered in different categories. These are:

### High-level Waste (HLW)

Radioactive waste that is radioactive enough for the decay heat to significantly increases its temperature and the temperature of its surroundings. This means that heat generation has to be taken into account when designing storage and disposal facilities.

This category of waste includes:

- The liquid residue that contains most of the radioactivity from the reprocessing of spent nuclear fuel,
- This material following solidification,
- Spent fuel (if it is declared a waste),
- Any other waste with similar radiological characteristics.

### Intermediate-level Waste (ILW)

Intermediate-level waste has radioactivity levels that are higher than low-level waste but which do not generate enough heat to require special storage or disposal facilities. However, like other radioactive waste it still needs to be contained to protect workers from the radiation. ILW arises mainly from the reprocessing of spent fuel and from general operations and maintenance at nuclear sites, and can include metal items such as fuel cladding and reactor components, graphite from reactor cores, and sludges from the treatment of radioactive liquid effluents.

### Low-level Waste (LLW)

Most LLW today arises from the operation of nuclear power stations and nuclear fuel reprocessing facilities, as well as the decommissioning and clean up of nuclear sites. Operational LLW is principally lightly contaminated miscellaneous waste arising from maintenance and monitoring, such as plastic, paper and metal. LLW from decommissioning is mainly soil, building materials and metal plant and equipment. Most LLW from nuclear licensed sites is currently disposed of at the Low-Level Waste Repository (LLWR) near Drigg in Cumbria.

### Very Low-level Waste (VLLW)

VLLW is a subset of LLW and falls into two distinct categories:

- Low-volume VLLW ('dustbin disposal'): Radioactive waste that can be safely disposed of to an unspecified destination with municipal, commercial or industrial waste. The radioactive risk from such material is low enough that controls on disposal of this material, after removal from the premises where the wastes arose, are not necessary.

- High-volume VLLW ('bulk disposal'): Radioactive waste that can be disposed of to specified landfill sites. After the waste is removed from its site of origin, it will be subject to controls on its disposal, which will be specified by the environmental regulators.

## Exempt Waste

Some radioactive waste is exempted from regulation by an Exemption Order issued under the Radioactive Substances Act 1993 (RSA93). L1 Exempt waste does not need an authorisation for disposal.

Radioactive waste management involves two fundamental approaches: the radioactive materials can be either released or discharged to the environment, or they must be confined and isolated from the biosphere until the noxious radionuclides have decayed to innocuous concentrations.

Releases of radioactivity to the environment generally occur as liquid or gaseous discharges (effluents) from nuclear facilities. The amount of radioactivity which can be released is based on allowable exposures to population groups and is controlled by national regulations and guidelines, usually based on recommendations of the International Committee on Radiation Protection.

There are very few instances where radioactive effluents from nuclear facilities can be released without some form of control or treatment to remove excessive radioactivity. Most of the radioactive wastes arising from the operation of nuclear fuel cycle facilities require processing to concentrate the radioactive elements into a smaller volume which can be more conveniently handled, thereby permitting the release or disposal of the bulk of the processed material.

Radioactive waste management involves the use of normal industrial operations and techniques adapted to cope with the barriers needed to protect the workers and operations from excessive radiation and contact with the radionuclides. Consequently, very often the equipment and its operating procedures become rather sophisticated to permit operation and maintenance behind the protective barriers. Actually, most research and development activities involving technology for the management of radioactive wastes are concerned with adapting normal industrial operations or laboratory techniques to the handling of radioactive materials.

It is the consensus amongst those involved with, or knowledgeable about, radioactive waste management, that technology and flow schemes have now been developed to the extent that they are workable. The basic knowledge required to collect, treat, package and store safely all nuclear fuel cycle wastes of concern is available. Furthermore, in many cases, optional technology exists for different situations or economic optimization.

Consequently, from a technical standpoint, it seems that lack of appropriate methods and technology for handling the radioactive waste need not become a bottleneck for the implementation of nuclear power. On the other hand, some of the required radioactive waste management technology still remains in its developmental stage. Much remains to be done in working out the engineering and design details, satisfying appropriate regulatory requirements, and adapting the technology to actual operating conditions and controls. These last remarks apply principally to those radioactive wastes arising from the reprocessing of irradiated fuel which essentially contain all the radioactive products of nuclear fission.

## Mill Tailings

Uranium mill tailings do not present a serious radiation hazard in open, well-ventilated areas. The principal concern involves the possiblity of hazardous concentrations of radon gas from the decay of the relatively small amounts of radium-226 in the tailings.

Tailings disposal areas can be appropriately located, managed and isolated. The technology exists to stabilize these tailings with a ground cover and vegetation and, if necessary, to protect them in other ways, in order to prevent their disperson by wind, water and man. Furthermore, such tailings areas can (and should be) recorded with the local and state land authorities, with restrictions in perpetuity being placed on their use.

## Effluents from Uranium Refining and Enrichment

In uranium refining and enrichment operations, the toxic, non-radioactive chemicals in the liquid and gaseous effluents are more of a problem than the radioactive contaminants. There are proven techniques and industrial practices available for preventing environmental pollution with the organic solvents, nitrates, sulfates, ammonia, chlorides, fluorides, fluorine and other chemicals used in the refining and enrichment processes. The depleted uranium from enrichment plants is currently not considered a waste and is recovered and stored.

However, the front end equipment of both uranium refineries and the processes for converting the refined uranium to uranium hexafluoride do eventually become radioactive from a build-up of the traces of radioactive elements still present in the concentrates and refined uranium. These radionuclides adsorp or plate on equipment surfaces and especially concentrate in the "ash" remaining from the uranium hexafluoride process. Such radioactive equipment and material generally are disposed of by suitably controlled land burial. Ocean dumping also has been used.

## Uranium Fuel Fabrication

The main waste management concern in the fabrication of uranium fuels comes from the non-radioactive elements and chemicals used in their manufacture. Uranium in waste is recovered and concentrations of uranium in effluents form uranium fuel fabrication plants are minimal. Normally, industrial non-radioactive waste management techniques and safety practices are used in uranium fuel fabrication and, generally, this is all that is needed. Uranium contaminated equipment can be readily decontaminated for safe disposal or possible reuse.

## Wastes within the Nuclear Fuel Cycle

The radioactivity induced by neutron capture is the major source of radioactive waste that requires management at the power reactor site. This neutron capture occurs in the corrosion products and other impurities in the coolant circulating through the reactor core, and in the structural components of the reactor that are exposed to high radiation levels. In addition, small quantities of radioactive fission products occur in the reactor coolant and fuel storage basin water as a result of an occasional fuel cladding failure.

The radioactive contaminants of main concern are fission products resulting from nuclear fission in the fuel elements. These fission products are released at the fuel reprocessing plant, where the fuel cladding is either chemically or mechanically breached and the fuel is dissolved. The soluble fission products dissolve into an aqueous solution, along with the Plutonium and unburned uranium. Gaseous fission products also are released to the plant off-gas systems during the breaching and dissolving operations.

The radionuclides remaining in the solution after recovery of the uranium and plutonium are the fission products and the actinides, also called the transuranics, which include the unrecovered plutonium and the heavier elements (Np, Am, Cm, etc) of the actinide series formed by neutron capture. Since these transuranics decay through alpha emission, the collective term alpha-emitters is commonly used. Waste contaminated with the alphaemitters is called alpha-contaminated or alpha-bearing waste.

The fission products and alpha-emitters contaminate all materials with which they come in contact and this contamination is passed along the contact chain. Hence, the formation of radioactive waste from what otherwise would be normal industrial waste. The extent to which such contaminated materials can be suitably decontaminated for some purpose only results in additional radioactive waste arising from the decontamination process. The materials handling and fabrication involving liquid and solid forms of fission products and transuranics, especially for plutonium in the fabrication of plutonium and uraniumplutonium oxide fuels, results in contaminated equipment and ventilation systems, all to be dealt with.

## Gaseous Wastes

The gaseous effluents from reactor operations may contain radioisotopes of the noble gases (principally, argon-41, krypton-85 and xenon-133), the radioiodines-129 and -131, tritium and oxides of carbon-14. However, as mentioned, the main release of gaseous radionuclides occurs when the spent nuclear fuel is breached and dissolved at the fuel reprocessing plant. There, the major releases are krypton-85, xenon-133, the radioiodines-129 and-131 and tritium. Xenon-133 with its relatively short (5.3 day) half-life is not anticipated to be a problem, with methods available for delay and decay, if necessary, prior to its release. The radioiodines can be eliminated from gaseous waste streams by counter-current scrubbing of the gas with aqueous solutions of caustic, mercuric nitrate, nitric acid, or by chemical adsorption on zeolites treated with silver or on other metals with an affinity for iodine. Another method is by adsorption on charcoal, which also can be impregnated with chemicals that will take up the iodine complexes.

While the extent to which tritium, krypton-85 and carbon-14 can be released into the atmosphere without becoming a matter of global concern has not been definitely established, analyses have indicated that this should not be a problem before the end of the century. Nevertheless, cryogenic (low temperature) separation methods are becoming available for removal of krypton-85 and xenon-133, if necessary. Other removal techniques, such as absorption in liquid fluorocarbons, adsorption on charcoal or other solids, and diffusion through membranes are under development.

far, no practical process has been found for the removal of tritium from waste effluents since f the tritium that is released is highly diluted as tritiated water. A promising approach for oval in fuel reprocessing is to confine, oxidize and collect the gaseous effluents from d other reprocessing plant head-end operations, converting the tritium (T2) to T2O tritium concentration of 1 to 5%. Other approaches being investigated are the adsorption of tritium on vanadium oxide, and isotopic separation processes such as high pressure and temperature diffusion through palladium films and the use of molecular sieves. Selective solvent extraction techniques also are being considered.

Carbon-14 can be removed in the oxide form from gaseous effluents by caustic scrubbing or adsorption on solid alkaline media, if the need arises. Radioactive particulate matter and aerosols in gaseous effluents can be removed by high efficiency filtration techniques.

## Liquid Wastes

Radioactive liquid wastes are loosely classified as high-level, intermediate-or medium-level and low-level, based primarily on their radioactivity levels. High-level waste has been generally accepted as the raffinate (liquid effluent) from the first cycle of fuel reprocessing operations. The raffinate contains over 99.9% of the non-gaseous fission products, the unrecovered plutonium and the higher actinides (transuranics). The distinction between medium- and low-level wastes is rather arbitrary and, in addition to their radionuclide content, the classification generally depends upon the extent to which they must be isolated and/or processed as compared with being released to the local surroundings. While medium- and low-level liquid wastes vary considerably with respect to their transuranic and fission product content, nevertheless there is a distinct difference between their treatment processes and those for high-level waste. Consequently they can be considered on the basis of this distinction.

As mentioned, high-level liquid waste is essentially the fuel reprocessing, first cycle raffinates. High-level liquid waste also can include concentrates of fission products and transuranics resulting from the processing of medium- and low-level liquid wastes.

High-level liquid waste is stored in suitably designed, cooled, stainless steel tanks, generally in steel-lined vaults. This method is considered safe for long periods although it requires close surveillance. However, it is the general opinion that these wastes should be converted to a solid form for long-term isolation from the terrestrial biosphere.

Several approaches have been developed for the solidification of high-level waste, including the use of fluidized beds, stirred beds, rotary kilns, sprays and pots. They all essentially involve heating the waste to between 400° and 1200°C, driving off all volatile constituents and leaving a calcined solid. However, most calcined wastes are relatively soluble and usually not considered a suitable product for disposal. Consequently, in most cases, borosilicate or phosphate glass forming constituents either are put into the calcining stage or are added as a second stage of the solidification process, incorporating the calcined waste into a glass melt, which cools to a vitrified product with leaching rates similar to those of Pyrex glass. There are at least 15 approaches to producing essentially the same vitrified product. However, with the exception of the French AVM process, which is scheduled to start full scale industrial operations at Marcoule in 1977, all the other processes have not been tested beyond the developmental stage. A novel approach under development at Eurochemic incorporates the radioactive calcine or glass into metal matrices.

Intermediate-level liquid wastes generally are characterized by smaller volumes and higher concentrations of radionuclides as compared to low-level liquid wastes. However, as indicated, there is no universally accepted distinction between the two. Intermediate-level liquid wastes can arise from spills, off-gas scrubber solutions, ion-exchange regenerates, cask and plant decontamination solutions, plant solvent washes, chemical decladding solutions, the reffinates from uranium and plutonium purification cycles and some laboratory wastes. Examples of low-level liquid wastes are evaporator and concentrator condensates, laundry wastes, condensates from process vessel ventilation systems and possible radionuclide leakage into cooling water or process steam.

All water soluble radioactivity released during reactor operations is introduced originally into either the primary coolant or the fuel storage basin water, mostly from fuel cladding failures. Radioactive particulate products of corrosion also are present. This radioactivity then spreads to other parts of the reactor system by leakage or treatment of these streams; for example, ion exchange regenerates and evaporator slurries. Intermediate- and low-level liquid waste streams arising from fuel reprocessing have been enumerated in the preceding paragraph. Besides decontamination, scrubbing and washing wastes, a mixed oxide ($P_UO_2$ + $UO_2$) fuel fabrication plant also generally has liquid waste streams from processing scrap material for plutonium recovery. These streams can be considered either intermediateor low-level material, depending on their characteristic actinide content.

Basic treatment techniques available to reduce levels of radioactivity in liquid waste streams are filtration and centrifugation (to eliminate radioactive particulates), evaporation, ion exchange, flocculation, and precipitation. Seepage ponds also can be used for liquid wastes containing very low concentrations of the shorter-lived fission products. Membrane concentration techniques are being examined but not yet generally used. The products of the foregoing basic treatment techniques are liquid concentrates and sludges which are "conditioned" to a solid form for either storage or disposal. Incineration and steam distillation techniques are available for the treatment of radioactive organic oils and solvents. Also, it is possible to reduce them to a solid state by the use of certain oil-absorbing solid materials.

## Solid Wastes

Besides the solid equipment and materials contaminated with either induced or contact radioactivity, solid radioactive wastes also include the solidified high-level liquid waste, "conditioned" concentrates from intermediate- and low-level liquid waste, loaded or spent ion exchange resins, and gaseous radionuclides incorporated into some solid medium for storage or disposal. Solid wastes fall into one of two categories, combustible or noncombustible. Depending on the facility, combustible and noncombustible waste may be mixed, and may require sorting prior to treatment and/or packaging for storage or disposal.

Combustible solid radioactive waste consists of a large variety of materials such as paper, rags, absorbent cotton, plastic sheeting, protective clothing gloves, rubber shoes, wood, cardboard, organic ion exchange resins, filter aids, combustible high efficiency filter media, etc. Since much of these waste materials is collected as general trash, it generally is sorted prior to incineration. Glove box operations for plutonium purification and fabrication introduce rubber or plastic materials.

Dry, conventional incineration techniques, although successfully practised, still involve problems of total combustion and off-gas cleaning. Other incineration techniques under development involve agitated hearths, cyclone furnaces and fluidized beds. Wet oxidation techniques involving acid digestion, now under development, and pyrolytic processes appear to be promising. Also the incorporation and incineration of combustible waste in molten salt mixtures is being examined.

While failed equipment may offer more difficult problems became of their large size and high level of radiation, there also is the steady stream of miscellaneous noncombustible radioactive waste from each type of nuclear fuel cycle facility. While the primary constituent of noncombustible solid waste is metal, including fuel cladding, other materials such as glass and concrete are significant. Combustible materials, such as organic ion exchange resins, plastics, grease and floor sweepings are often included with noncombustible waste.

Typical large items of equipment from a fuel reprocessing plant may consist of dissolvers, solvent extraction columns or units and concentrators which might be up to 3 metres in diameter and 10 metres in height. Noncombustible waste from reactors can include large components from the reactor core. Many of the large equipment items must be dismantled, sometimes prior to removal from the system. The normal approach is to flush and decontaminate such equipment to the extent possible. Both fuel reprocessing plant and reactor equipment may be highly radioactive, requiring remote handling and personnel shielding equivalent to several feet of concrete. Such equipment may be stored until the radioactivity has decayed to lower levels.

Failed equipment from plutonium processing and mixed oxide fuel fabrication operations, such as glove boxes, may have low enough radiation levels to permit dismantling without shielding or with light shielding for the workers. However, the presence of plutonium requires elaborate precautions to prevent its spread to the environs.

Generally, noncombustible waste is packaged with appropriate shielding. Where possible, the waste can be crushed or melted for volume reduction. Smaller units of noncombustible wastes, such as incineration ash, ion-exchange resins, and sludges and residues from the treatment of intermediate- and low-level liquid wastes are incorporated in concrete and bitumen. Ureaformaldehyde resins, and salt matrices and other absorbants and packing materials also are being used.

## Storage

Storage implies easy retrievability. Storage facilities for radioactive wastes must prevent radiation exposure to man and provide physical confinement. Continuing surveillance is required to assure that these two conditions are being met.

It has not been necessary to store radioactive gaseous materials nor is it likely to be for some time to come. However, they can be compressed and stored in metal cylinders. Techniques are being developed to incorporate them within solid matrices and adsorbants as a protection against their sudden release should a container fail. Radioactive liquid wastes can be conveniently stored in tanks or other containers. Storage methods for radioactive solid wastes ranges from the use of concrete or earthen vaults for highly radioactive materials to simple fire-proof structures for low-levels of radiation. Underground mines and tunnels in dry geologic formations also are being utilized. The experience gained from storing irradiated fuel elements is applicable to the storage of solid high-level waste in water basins or air-cooled vaults.

## Disposal

Disposal implies a final resting place or repository with no intent of retrieving the material. Furthermore, once the safety and confinement of the disposal method has been assured, surveillance of the site can be minimal or even discontinued. It is generally accepted that, for disposal, it is better for all radioactive waste to be in a relatively insoluble solid form which is less susceptible to dispersion from a disposal site than a liquid or a gaseous form. Also, the method of packaging can be used as a barrier between the contents and its surroundings. On the other hand, the USSR has demonstrated the safe injection of intermediate- and low-level liquid wastes into porous formations in deep geologic structures. In the USA, Oak Ridge National Laboratory is demonstrating the injection of a mixture of liquid waste and solidifying agents into shale seams by hydrofracture.

Regardless of certain attractions for using outer space, disposal of radioactive solid wastes will be terrestrial for some time to come. Land burial and sea dumping are being utilized for the disposal of intermediate- and low-level solid wastes. Land burial techniques range from the use of concrete trenches and vaults, which may then be filled with concrete, to direct burial of the packaged waste in trenches, pits or landfill. Protection against migration of the radionuclides into ground-water is of paramount concern. Sea dumping requires suitable packaging and preparation to withstand or equalize the water pressure on the container and sufficient weight to allow it to sink to the ocean depths and remain there. The amount of radionuclides disposed of at any one site (land or sea) is carefully controlled so as to not-exceed the environmental capacity. The extent to which solid waste contaminated with the longer-lived alpha-emitters (alpha-bearing waste) can be disposed of at land burial or sea dumping sites still is under study which includes their concentrations in the waste as well as total quantities.

The disposal of suitably packaged low- and intermediate-level solid wastes in an underground salt formation is being demonstrated in the Federal Republic of Germany. Many countries now are considering a similar approach.

The activation product of most concern found in solid waste is cobalt-60 with a half-life of 5.3 years. Consequently, highly irradiated reactor components generally will have decayed to innocuous levels within 150—200 years. For this relatively short time period, land burial and properly packaged sea disposal would seem to be suitable.

On the other hand, wastes contaminated with the longer-lived fission products may require from 600 to 1000 years before they do not constitue a radiological hazard. Wastes contaminated with plutonium-239 (a half-life of 24,300 years) can require isolation periods of up to 500, 000 years. Other transuranic actinides can require more or less such geologic time periods before their concentrations can be considered tolerable.

Currently, there has been no method demonstrated for the disposal of the solidified highlevel and other wastes significantly contaminated with the alpha-emitters, possibly because there has been no need for such facilities. On the other hand, public and political concerns regarding this question can no longer be ignored. Many countries now are examining the possibilities of disposing of the solidified high-level and alpha-bearing wastes into suitable underground geologic structures (geologic disposal) which is the only option available with today's technology. Salt, clay and crystalline rock formations are being considered. One can expect that within the next five to seven years, several demonstration projects for geologic disposal will be operating. Disposal into the deep ocean sediments or into geologically stable areas of the sea bed has been proposed, using the technology developed recently by exploring and developing off-shore oil deposits. However, it will be some time (ten years or more) before this can be considered more than exploratory, assuming the technical requirements can be met economically.

An attractive idea is to remove the actinide alpha-emitters with the very long half-lives from the bulk of the wastes, especially from the high-level waste. The remaining fission product contaminated material then would need to be isolated up to only 1000 years, a more workable time period. Without the high heating generating fission products, the low heat generating actinides could be disposed of in some more suitable manner or recycled back to nuclear reactors for burning (transmutation) to shorter lived or stable elements. It is technically feasible to remove the actinides to negligible concentrations but uncertain if it is operationally obtainable. Furthermore, it will take several decades to demonstrate the feasibility of actinide recycle.

## References


- Hazardous-waste-management, technology: britannica.com, Retrieved 28 April, 2019
- Hazardous-waste-management, earth-and-planetary-sciences: sciencedirect.com, Retrieved 11 August, 2019
- The-management-of-hazardous-waste-in-developing-countries, management-of-hazardous-wastes: intechopen.com, Retrieved 24 June, 2019
- Fundamentals-of-the-management-of-radioactive-waste-101350: sepa.org.uk, Retrieved 17 July, 2019
- Magazines, bulletin, bull18-5/185-604644047: iaea.org, Retrieved 10 March, 2019

3

# Hazardous Waste Treatment Technologies

There are various methods and technologies which are used to treat hazardous waste. Pyrolysis, gastification, incineration, bioremediation and landfill are some of the different chemical, biological and physical treatment methods. This chapter discusses these techniques and methods related to hazardous waste treatment in detail.

The treatment technologies have been categorized as physical, chemical, biological, thermal, or stabilization/fixation.

Physical treatment processes include gravity separation, phase change systems, such as air and steam stripping of volatiles from liquid wastes, and various filtering operations, including carbon adsorption.

Chemical treatment transforms waste into less hazardous substances using such techniques as pH neutralisation, oxidation or reduction, and precipitation. Biological treatment uses microorganisms to degrade organic compounds in the waste stream.

Thermal destruction processes include incineration, which is increasingly becoming a preferred option for the treatment of hazardous wastes, and pyrolysis, which is the chemical decomposition of waste, is brought about by heating the material in the absence of oxygen.

Stabilisation techniques involve removal of excess of water from a waste and solidifying the remainder either by mixing it with a stabilising agent such as Portland cement, or vitrifying it to a glassy substance.

Most treatment measures have both physical and chemical aspects. The appropriate treatment technology for the hazardous wastes depends on the nature of the wastes. The type of physical treatment to be applied to wastes depends strongly upon the physical properties of the material treated, including the state of matter, solubility in water and organic solvents, density, volatility, boiling point and melting point.

## Physical Treatment of Hazardous Waste

Various physical treatment methods adopted in industry are: adsorption, resin adsorption, sedimentation, electro dialysis, reverse osmosis, solvent extraction, distillation, evaporation and filtration.

## Adsorption

Adsorption on activated carbon occurs when a molecule is brought up to its surface and held there by physical and chemical forces. This process is reversible, thus allowing activated carbon to be regenerated and reused by proper application of heat and steam, or solvent.

The factors that relate to adsorption capacity are:

- Greater surface area produces greater adsorption capacity e.g. Activated carbon has large surface area (500-1500 $m^2/g$).
- Adsorptivity increases as the solubility of the solute (in solvent) decreases. Thus, for hydrocarbons, adsorption increases with molecular weight.
- For solutes with ionisable groups, maximum adsorption will be achieved at a pH corresponding to minimum ionisation.
- Adsorption capacity decreases with increasing temperature.

One additional point to be noted is that biological activity usually takes place in a carbon bed. If the concentration of the adsorbed species is high enough and the material is biodegradable and non toxic to the bacteria, then biological activity may significantly increase the effective removal capacity. Removal through adsorption by activated carbon has been applied to non aqueous waste stream such as petroleum fraction, syrups, vegetable oils, and pharmaceutical preparations. Colour removal is the most common objective in such cases. Current waste treatment applications are limited to aqueous solutions.

## Resin Adsorption

Waste treatment by resin involves two basic steps:

- Contacting the liquid waste stream with resin and allowing the resin to absorb the solutes from the solution; and
- Subsequently regenerating the resins by removing the adsorbed chemicals, by simply washing with proper solvent.

The adsorption of a non-polar molecule on to a hydrophobic resin (e.g., styrene divinyl-benzene-based resin) results primarily from the effect of Vander Waal's forces. In other cases other type of interactions such as dipole-dipole interaction and hydrogen bonding are important. In a few cases ion-exchange mechanism may be involved. For the removal of organic dye wastes from water, two different resins were employed: In this case the waste stream is first contacted with a normal polymeric adsorbent and then with an ion exchange resin.

## Sedimentation

Sedimentation is a physical process whereby particles suspended in a liquid settle by means of gravity.

The fundamental elements of most sedimentation processes are:

- A basin or container of sufficient size to maintain the liquid to be treated in a relatively quiescent state for a specified period of time.

- A means of directing the liquid to be treated into the above basin in a manner conducive to settling.
- A means of physically removing the settled particles from the liquid.

Sedimentation can be either a batch or a continuous process. Continuous processes are by far the most common, particularly when large volumes of liquid are to be treated. This technique has been widely used in the removal of heavy metals from iron and steel industry waste water; removal of fluoride from aluminium production waste water; and removal of heavy metals from waste water from copper smelting and from metal finishing industry and waste water stream from organic chemicals.

## Electro-dialysis

The electro-dialysis involves the separation of an aqueous stream (more concentrated in electrolyte than the original) and a depleted stream. Success of the process depends on special synthetic membranes, usually based on ion-exchange resins, which are permeable only to a single type of ion. Cation exchange membranes permit passage only of positive ions, under the influence of electric field, while anion exchange membranes permit passage only of negatively charged ions.

The feed water is passed through compartments formed by the spaces between alternating cation-permeable and anion permeable membranes held in a stack. At each end of the stack is an electrode having the same area as the membranes. A dc potential applied across the stack causes the positive and negative ions to migrate in opposite directions.

This technique has been used for desalination to produce potable water from brackish well water. In food industry electro-dialysis is used for desalting whey. The chemical industry uses this technique for enriching or depleting solutions, and for removing mineral constituents from product streams.

## Reverse Osmosis

This technique which is most widely used consists of a membrane permeable to solvent but impermeable to most dissolved species, both organic and inorganic. These devices use pressure to force the contaminated water against the semipermeable membrane. The membrane acts as a filter, allowing the water to be pushed through the pores, but restricting the passage of larger molecules that are to be removed.

Cellulose acetate membranes were used in the past, but nowadays polysulphones and polyamides are increasingly popular for use at high pH values. Because of the susceptibility of the membranes to chemical attack and fouling, and the susceptibility of the flow system to plugging and erosion, it is common to preprocess the feed water to remove oxidising materials. The reverse osmosis technique has been widely used for desalination of sea or brackish water.

It has also been successfully used in the treatment of electroplating rinse waters, not only to meet effluent discharge standards, but also to recover concentrated metal salt solutions for reuse. It has also been used for treatment of waste stream from paper and food processing industries.

## Solvent Extraction

Solvent extraction is the separation of the constituents of a liquid solution by contact with another immiscible liquid. If the substances comprising the original solution distribute themselves differently between the two liquid phases, a certain degree of separation will result and this may be enhanced by the use of multiple contacts. The major application of solvent extraction to waste treatment has been in the removal of phenol from by-product water produced in coal coking, petroleum refining, and chemical synthesis that involve phenol.

The use of supercritical fluids (SCFs) most commonly $Co_2$ as extraction solvent, has been one of the more promising approaches to solvent extraction. SCFs are fluids existing at or above the lowest temperature at which condensation may occur. Above the critical temperature certain fluids exhibit characteristics that enhance their solvent properties.

Organic materials, which are only slightly soluble in particular solvents at room temperature, become completely miscible with the solvent when under supercritical conditions. The excellent solvent properties result from the rapid mass transfer ability and the very low density that characterises an SCF. Major advantages of SCFs are short residence times with no char formation.

Some of the important application of these SCFs, has been in the extraction of organ halide pesticide from soil, extraction of oil from emulsions used in aluminium and steel processing, and regeneration of spent activated carbon. Supercritical ethane has been used to purify waste oils contaminated with PCBs, metals and water.

## Distillation

Distillation is expensive and energy intensive and can probably be justified only in cases where valuable product recovery is feasible (e.g., solvent recovery). This technique has only limited application in the treatment of dilute aqueous hazardous wastes.

## Evaporation

Evaporation process is used for the treatment of hazardous waste such as radioactive liquids and sludges and concentrating of plating and paint solvent waste among many other applications. It is capable of handling liquids, slurries and sometimes sludges, both organic and inorganic, containing suspended or dissolved solids or dissolved liquids, where one of the components is essentially non volatile. It can be used to reduce waste volume prior to land fill disposal or incineration.

The major disadvantages of evaporation are high capital and operating costs and high energy requirements. This process is more adaptable to waste waters with high concentrations of pollutants.

## Filtration

Filtration is well-developed economical process used in the full scale treatment of many industrial waste waters and waste sludges. Energy requirements are relatively low, and operational parameters are well defined. However it is not a primary treatment process and is often used in conjunction with precipitation, flocculation, and sedimentation to remove these solids.

## Flocculation

The various phenomena that occur during flocculation can be grouped in to two sequential mechanisms. Chemically induced destabilisation of repulsive surface related forces, thus allowing particles to stick together when they touch and Chemical bridging and physical enmeshment between the non repelling particles, allowing for the formation of large particles.

Chemicals used for flocculation include alum, lime, ferric chloride, ferrous sulphate and poly electrolytes. Poly electrolytes consist of long chain, water soluble polymers such as polyacrylamides. They are used either in conjunction with inorganic flocculants, or as primary flocculating agent. The inorganic flocculants such as alum, upon mixing with water, the slightly higher pH of water causes them to hydrolyse to form gelatinous precipitates of aluminium hydroxide.

It is partially because of their large surface area, they are able to enmesh small particles, and thereby create larger particles. Once suspended particles have been flocculated into larger particles, they usually can be removed from the liquid by sedimentation, provided that a sufficient density difference exists between the suspended matter and the liquid.

## Physical Methods of Hazardous Wastewater Treatment

Hazardous waste comprises all types of waste with the potential to cause a harmful effect on the environment and pet and human health. It is generated from multiple sources, including industries, commercial properties and households and comes in solid, liquid and gaseous forms.

There are different local and state laws regarding the management of hazardous waste in different localities. Irrespective of your jurisdiction, the management starts from a proper hazardous waste collection from your Utah property through to its eventual disposal.

There are many methods of waste treatment after its collection using the appropriate structures recommended by environmental protection authorities. One of the most common and inexpensive ones is physical treatment. The following are the physical treatment options for hazardous wastewater:

## Sedimentation

In this treatment technique, the waste is separated into a liquid and a solid. The solid waste particles in the liquid are left to settle at a container's bottom through gravity. Sedimentation is done in a continuous or batch process.

Continuous sedimentation is the standard option and generally used for the treatment for large quantities of liquid waste. It is often used in the separation of heavy metals in the steel, copper and iron industries and fluoride in the aluminum industry.

## Electro-dialysis

This treatment method comprises the separation of wastewater into a depleted and aqueous stream. The wastewater passes through alternating cation and anion-permeable membranes in a compartment.

A direct current is then applied to allow the passage of cations and anions to opposite directions. This results in solutions with elevated concentrations of positive and negative ions and another with a low ion concentration.

Electro-dialysis is used to enrich or deplete chemical solutions in manufacturing, desalting whey in the food sector and generating potable water from saline water.

## Reverse Osmosis

This uses a semi-permeable membrane for the separation of dissolved organic and inorganic elements in wastewater. The wastewater is forced through the semi-permeable membrane by pressure, and larger molecules are filtered out by the small membrane pores.

Polyamide membranes have largely replaced polysulphone ones for wastewater treatment nowadays owing to their ability to withstand liquids with high pH. Reverse osmosis is usually used in the desalinization of brackish water and treating electroplating rinse waters.

## Solvent Extraction

This involves the separation of the components of a liquid through contact with an immiscible liquid. The most common solvent used in the treatment technique is supercritical fluid (SCF) mainly $CO_2$.

These fluids exist at the lowest temperature where condensation occurs and have a low density and fast mass ion transfer when mixed with other liquids. Solvent extraction is used for extracting oil from the emulsions used in steel and aluminum processing and organ halide pesticide from treated soil.

Superficial ethane as a solvent is also useful for the purification of waste oils contaminated with water, metals, and PCBs.

Some companies and household have tried handling their hazardous wastewater to minimize costs. This, in most cases, puts their employees at risk since the "treated" water is still often dangerous to human health, the environment and, their machines.

The physical processes above sometimes used with chemical treatment techniques are the guaranteed options for truly safe wastewater.

## Chemical Treatment of Hazardous Waste

Chemical treatment (of hazardous waste) refers to the treatment methods that are used to effect the complete breakdown of hazardous waste into non-toxic gases or, more frequently, to modify the chemical properties of the waste, for example, through reduction of water solubility or neutralisation of acidity or alkalinity.

### Pyrolysis

Pyrolysis is a form of treatment that chemically decomposes organic materials by heat in the absence of oxygen. Pyrolysis typically occurs under pressure and at operating temperatures above 430 ¡C (800 ¡F). In practice, it is not possible to achieve a completely oxygen-free atmosphere. Because some oxygen is present in any pyrolysis system, a small amount of oxidation occurs. If volatile or semi-volatile materials are present in the waste, thermal desorption will also occur.

Organic materials are transformed into gases, small quantities of liquid, and a solid residue containing carbon and ash. The off-gases may also be treated in a secondary thermal oxidation unit. Particulate removal equipment is also required. Several types of pyrolysis units are available, including the rotary kiln, rotary hearth furnace, and fluidized bed furnace. These units are similar to incinerators except that they operate at lower temperatures and with less air supply.

A molten salt process may also be used for waste pyrolysis. In molten-salt oxidation (MSO), combustible waste is oxidized in a bath of molten salts (at 500–950¡C). There is no direct flame, and this prevents many of the problems associated with incineration. Shredded solid waste is injected with air under the surface of a molten salt bath. Hot gases rise through the molten salt bath. The salt, being alkaline, scrubs acids from the gases. The heat of the molten salt degrades and melts the waste material. Because the salt bath is liquid, it also removes some particles in the gas. By-products are retained in the melt. Gases exiting the salt bath are treated before discharge to the atmosphere. Spent molten salt is tapped from the reactor, cooled, and placed in a landfill.

### Limitations and Concerns

The technology requires drying of soil prior to treatment.

Limited performance data are available for systems treating hazardous wastes containing polychlorinated biphenyls (PCBs), dioxins, and other organics. There is concern that systems that destroy chlorinated organic molecules by heat have the potential to create products of incomplete combustion, including dioxins and furans. These compounds are extremely toxic in the parts per trillion range. The MSO process reportedly does not produce dioxins and furans.

The molten salt is usually recycled in the reactor chamber. However, depending on the waste treated (especially inorganics) and the amount of ash, spent molten salt may be hazardous and require special care in disposal.

Pyrolysis is not effective in either destroying or physically separating inorganics from the contaminated medium. Volatile metals may be removed as a result of the higher temperatures associated with the process, but they are not destroyed. Byproducts containing heavy metals may require stabilization before final disposal.

When the off-gases are cooled, liquids condense, producing an oil/tar residue and contaminated water. These oils and tars may be hazardous wastes, requiring proper treatment, storage, and disposal.

## Applicability

Pyrolysis treats and destroys semi-volatile organic compounds (SVOCs), fuels, and pesticides in soil. The process is applicable for the treatment of organics from refinery wastes, coal tar wastes, creosote-contaminated soils, hydrocarbons, and volatile organic compounds (VOCs). MSO is also being tested for treating mixed radioactive wastes.

## Technology Development Status

Pyrolysis is a developed technology. The basic concepts of the process have been validated. However, there are limited performance data on the treatment of hazardous wastes containing PCBs, dioxins, and other organics.

## Gasification

Gasification and pyrolysis are similar processes; both decompose organic waste by exposing it to high temperatures. Both processes limit the amount of oxygen present during decomposition; gasification allows a small amount of oxygen, pyrolysis allows none. In other words, gasification and pyrolysis limit or prevent oxidation.

In the presence of air, heat causes organic materials to burn. Burning or oxidation is what typical incinerators do. The burning of waste in incinerators causes well-known negative environmental and public health effects. Incinerators emit nitrogen oxides, sulfur dioxide, particulate matter, carbon monoxide, carbon dioxide, acid gases, lead, cadmium and mercury, and organic compounds, such as dioxins and furans, into the atmosphere. In 1960 mass burn incinerators burned 30% of the municipal solid waste in the United States. By 1988 this total had dropped to 13% because of air pollution problems.

Gasification facilities produce gas—primarily carbon monoxide and hydrogen (85%)— plus hydrocarbon oils, char and ash. Gasification plants' air emissions also include nitrogen oxides, sulfur dioxide, particulate matter, carbon monoxide, carbon dioxide, methane, hydrogen chloride, hydrogen fluoride, ammonia, heavy metals mercury and cadmium, dioxins and furans.

Gasification of municipal solid waste, household garbage and commercial waste products was used in the United States in the 1970s, but those plants were closed because of operating and financial problems. Today there are only a handful of pyrolysis units burning municipal solid waste, located in Japan, Taiwan, Great Britain and Canada.

Gasification facilities share the same environmental problems associated with mass burn incinerators including:

- Air pollution.
- Water pollution.
- Disposal of ash and other by-products.
- Large amounts of water for cooling purposes.
- Health, safety, and odor impacts.
- Disincentives for waste reduction.
- Diversion of waste from composting and recycling.

The gas produced from municipal solid waste contains hazardous organic compounds but the technology to remove these toxins does not exist on a commercial scale.

The ash which remains after gasification, 8% to 15% of the original volume, is toxic and presents special problems because of the acidic, or low pH, conditions in landfills. Leaching of toxic metals cadmium, lead, and mercury occurs more rapidly at low pH, resulting in contaminated groundwater.

The organic material in solid waste necessary for the gasification process is the part of the waste stream most valuable for composting. Sixty percent of solid waste is paper, cardboard, wood, yard waste, and food scraps.

The economics of gas transport require power generation units burning the gas to locate at or near gasification facilities; multiple smoke stacks in an area bring larger air pollution problems.

Gasification undermines waste composting and recycling programs. Technologies which utilize high temperatures to process solid waste incineration, waste-to-energy, gasification, and others have negative effects on low tech methods, a problem which is compounded when recyclable materials are not removed before processing. For example, paper is a valuable material which is lost when reduced to a gas. Aluminum, steel, and glass are easily recycled when separated from solid waste but unrecoverable from gasification process slag.

The principal benefit touted by promoters of high temperature technologies, energy recovery, is far outweighed by the energy loss tied to the destruction of energy-intensive solid waste materials. Gasification produces a low-Btu gas with 25% of the heat content of natural gas. But recycling glass, aluminum, and paper preserves the large amounts of energy needed to produce them from silica, bauxite, and trees; products made from recycled cullet, cans, and paper utilize a fraction of the energy needed to make them from raw materials.

## The Gasification Process

Gasification is a process that chemically and physically changes biomass through the addition of heat in an oxygen-starved environment. The end products of gasification include solids, ash and slag, liquids and synthesis gas, or syngas. The gas has a calorific value, or potential heat content, equivalent to 25% that of natural gas if ambient air is used or 40% if oxygen-enriched air is used. Figure on the following page illustrates the process: combustion chambers, power boiler, pollution controls, and pollution by-products: bottom ash and slag, fly ash, and air emissions.

A waste industry trade association monograph describes the starved air combustion process.

This type of incineration consists of two chambers: the primary is operated at below the stochiometric air requirement and the second operated under excess air conditions. The waste is fed into the primary chamber and semi-pyrolysed, releasing moisture and volatile components. The heat is provided by the controlled combustion of fixed carbon within the waste. The syngas that is driven off contains a high calorific value and can act as a feedstock for the secondary chamber. Importantly, combustion air is then added to the syngas making it highly combustible and prone to self-ignition. The secondary chamber is equipped with a conventional burner to maintain operating temperature at all times. The combined gases are combusted in the secondary chamber.

The industry monograph describes pyrolysis as a similar process but with no oxygen present in the first combustion chamber: "Materials are heated in the absence of Oxygen to about 800 °C. Hydrocarbons are converted to simple gases leaving a residue of carbon char, inert materials and heavy metals." Plasma arc furnaces replace the fossil-fuel heat source with an electric arc.

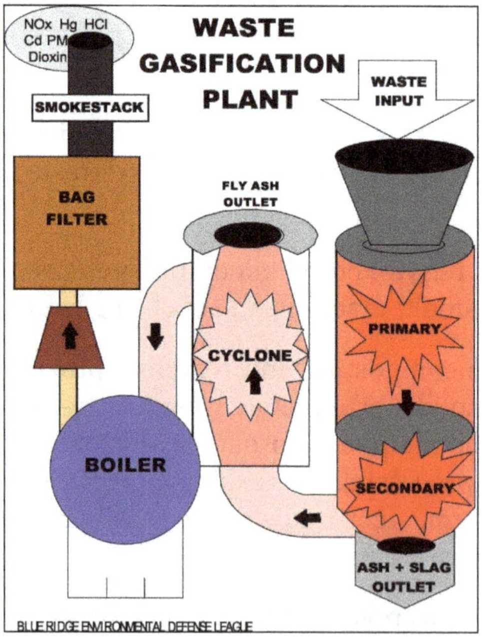

Waste gasification process.

Although some gasification facilities have been designed and constructed in the past two decades, most have been demonstration and laboratory-scale systems. A few large scale demonstration plants in the US experienced technological problems and are no longer operating. In 2006 the US Environmental Protection Agency Office Air and Radiation selected a solid waste gasification project for research and development, but the project has not moved forward. There are currently no commercial-scale solid waste gasification systems operating in the United States.

## Emission of Air Pollutants

Incineration of solid waste is a technology which is being phased out across the United States. According to Dr. Paul Connett, "Since 1985 over 280 incinerator proposals have either been defeated outright or put on hold." Unreliability, economics, environmental concerns, and citizen opposition have forced municipal officials to find new methods of managing solid waste.

Gasification shares many characteristics with incineration. At high temperatures used in incineration and gasification, toxic metals including cadmium and mercury, acid gases including hydrochloric acid, and ozone-forming nitrogen oxides are released. Also, dioxins and furans are created in the cooling process following the burning of ordinary paper and plastic. These poisons are dangerous at extremely low levels and modern pollution control devices do a poor job of reducing these emissions into the atmosphere. Some including mercury and dioxin are persistent and bioaccumulative; they resist breakdown in the environment and are concentrated in the food chain.

The Environmental Protection Agency database on industrial air pollution—Emission Factors AP-42—does not include a specific category for solid waste gasification. However, there is a category for "starved-air" combustion. When the oxygen level in a heating unit is reduced to a level below that needed for complete combustion, it is called a starved-air unit. EPA does list Emission Factors for Modular Starved-Air Combustors. The data table is reproduced in Appendix C of this report.

The regulatory requirements and permitting decisions for gasification units would likely be based on the facility which is most similar, a starved-air combustor. The AP-42 includes emission levels for 13 pollutants, with and without electrostatic precipitators: (beginning with the highest) carbon dioxide, particulates, sulfur dioxide, nitrogen oxides, hydrochloric acid, carbon monoxide, mercury, nickel, chromium, lead, cadmium, arsenic, and chlorinated dibenzo-furans and -dioxins. The AP-42 factors show that emissions of only five of the pollutants are reduced by ESPs, six are unchanged, and dioxins and furans are higher. The EPA AP-42 lists air emissions in pounds per ton and kilograms per metric tonne of waste input.

Using the AP-42, one can calculate the emissions from an air pollution source. For example, a 100 ton-per-day starved-air combustor could burn 36,500 tons of garbage per year. Based on these data, annual air emissions would be as shown in table.

Table: Air Pollutants from Starved-air Combustion/Gasification.

| Pollutant | Uncontrolled | Electrostatic precipitator |
|---|---|---|
| Particulate matter | 125,195 pounds | 12,702 pounds |
| Sulfur dioxide | 117,895 pounds | * |
| Nitrogen oxides | 115,340 pounds | * |
| Hydrochloric acid | 78,475 pounds | * |
| Carbon monoxide | 10,913 pounds | * |
| Mercury | 204 pounds | * |

| Nickel | 201 pounds | 37 pounds |
|---|---|---|
| Chromium | 121 pounds | 22 pounds |
| Lead | 103 pounds | - |
| Cadmium | 88 pounds | 17 pounds |
| Arsenic | 24 pounds | 4 pounds |
| Dioxins/furans | 0.11 pound | 0.14 pound |

* same as uncontrolled

The EPA's AP-42 Emission Factors are the basis for air pollution regulations in all 50 states. There are no other regulatory emissions data or emissions factors for solid waste combustors. With no new federal studies to demonstrate otherwise, allowable hazardous air pollutant emissions from permitted gasification units will be identical to the limits for incinerators.

## Waste Gasification Adds Greenhouse Gases

In addition to the currently regulated air pollutants from municipal solid waste gasification, the process also adds greenhouse gases to the atmosphere. The biggest contributor to global warming, carbon dioxide, could soon be regulated as a pollutant by the US Environmental Protection Agency.

Using municipal solid waste for fuel releases into the atmosphere the carbon which is in the paper, cardboard, food wastes, yard wastes and other biological materials, plus the carbon in plastic products and containers made from petroleum. The gasification of petroleum-based plastics adds to greenhouse gases in the same way as burning fossil fuels such as coal, oil or natural gas.

It is the release of carbon, carbon that has been locked up in fossil deposits for millions of years, that is driving global warming. However, carbon released into the atmosphere from the burning or gasification of waste products made from trees and crops also adds to greenhouse gas levels in the atmosphere. Over time some of this carbon is taken up again by new growth in forests, but in the short-term the damage is done. The US EPA published an analysis of greenhouse gas (GHG) impacts caused by various solid waste management options. As can be seen in the EPA illustration in figure, solid waste disposal plays a role in global warming including the greenhouse gases from waste combustion.

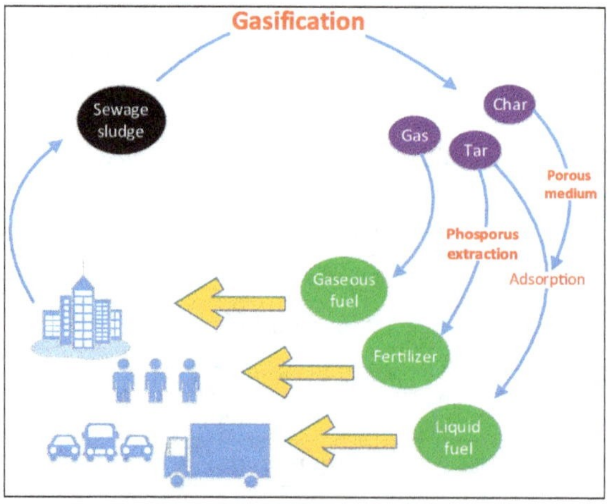


Greenhouse gas sources.

The different sources of greenhouse gas (GHG) emissions from waste are shown above. The disposal of solid waste produces GHGs in a number of ways. First, the anaerobic decomposition of waste in landfills produces methane, a GHG 21 times more potent than carbon dioxide. Second, the incineration of waste produces carbon dioxide as a by-product. In addition, the transportation of waste to disposal sites produces GHGs from the combustion of the fuel used in the equipment. Finally, the disposal of materials indicate that new products are being produced as replacements; this production often requires the use of fossil fuels to obtain raw materials and manufacture the items.

Recycling and composting far exceed all other waste management options in reducing greenhouse gas emissions. A study by the Tellus Institute for the State of Massachusetts concluded that:

> "Our review of the LCA (life cycle analysis) literature and our modeling outputs confirm that, after source reduction, waste diversion through recycling and composting is the most advantageous management option from an environmental and energy perspective. Because it releases bound carbon in materials such as plastics, thermal conversion of certain materials to fuels or energy is problematic from a climate change perspective even at the potentially high energy recovery levels of advanced conversion technologies. Similar to the situation for WTE (waste to energy) incinerators, the capital requirements for building alternative technology facilities and their likely need for long-term contracts to ensure an adequate feedstock waste stream may limit the future flexibility of the state's overall materials management efforts".

Waste gasification adds both directly and indirectly to the build-up of greenhouse gases in the earth's atmosphere. It is not a solution to the global warming problem; it is instead part of the problem-a part that should be eliminated as a waste management option.

> "By destroying resources rather than conserving them, all incinerators, including mass burn, pyrolysis, plasma and gasification, cause significant and unnecessary lifecycle GHG emissions".

## Emission of Water Pollutants

Combustion and gasification may reduce solid waste volume by 85% to 92%, but the remaining ash, 8-15% of the original waste, must be disposed of. Typically, it ends up in landfills and containssome of the same toxic compounds listed above, particularly heavy metals. Toxics in combustion ash ultimately find their way to groundwater and surface water, contaminating water supplies.

A national controversy about ash toxicity erupted in 1995 when then-EPA Administrator Carol Browner allowed incinerator operators to mix bottom ash and fly ash together prior to toxicity testing. Fly ash raises the pH of the ash, reducing the reliability of the tests. But citizens who gathered samples of ash from incinerators which had passed the EPA's tests found very high levels of toxic metals. The results of those tests are compiled in table, below. Dr. Paul Connett described the methods used by landfill neighbors to get laboratory samples.

The ash from Ogden Martin's 990 ton-per-day MSW incinerator in Syracuse, NY, which went on line in November 1994, was classified as non-hazardous after it passed the TCLP toxicity characteristic leaching procedure test. Citizens living close to the landfill, where the ash is used as a daily cover, surreptitiously retrieved some of this ash and asked N.Y. State Wildlife Pathologist, Ward

Stone, to test the ash for heavy metals, because the landfill abuts the Montezuma Wildlife Refuge in Seneca County, NY. Stone analyzed the ash for total heavy metal content. The ash was found to contain hazardous levels of lead, cadmium and mercury.

Table: Heavy metals in incinerator ash compared to soil.

| Metal Tested Results in parts per Million (ppm) | Ogden Martin Syracuse, NY Incinerator | Foster Wheeler Hudson Falls, NY incinerator | Mean background levels in US soils |
|---|---|---|---|
| LEAD | 1400 ppm | 2650 ppm | 35 ppm |
| CADMIUM | 40.1 ppm | 60.3 ppm | 0.30 ppm |
| MERCURY | 4.3 ppm | 4.1 ppm | 0.18 ppm |

The TCLP test is still used to determine toxicity of solid waste ash; passing the test allows it to be disposed of in a solid waste landfill. However, passing the test does not mean that the waste ash is not hazardous. As the tests done by Ward Stone proved, the ash residue left after waste is burned may contain hundreds of times as much toxic metal as uncontaminated soil. Dr. Connett summarized the environmental and public health impacts:

> "Carol Browner, the head of the US EPA, handed the incinerator industry a huge gift by allowing the incinerator operators to mix the fly ash with the bottom ash prior to testing. The bottom ash contains no lime. A TCLP test applied to this ash alone would probably yield a pH in the acid range and one would anticipate a failure rate of about 30 to 40 percent of the time, as in the case of the old EP Toxicity Tests. However, when the fly ash is mixed with the bottom ash, the lime protects the bottom ash as well. Again, the leaching medium will not reach pH 5, but stay in the range of lead's least solubility. As bottom ash represents 80 to 90 percent of the total ash, this protection represents a huge financial bonanza to the incineration industry. While the TCLP test serves the industry's interests, it presents a major threat to human health and the environment".

Gasification units produce both bottom ash and fly ash. The toxicity of gasification combustor ash would be no different than incinerator ash because the source, municipal solid waste, is the same.

## Diversion of Wastes from Recycling and Composting

Zero Waste is a principle which maximizes recycling, minimizes waste, and reduces consumption. Zero waste ensures that products are made to be reused or recycled back into nature or the marketplace and tackles the fundamental issues of corporate irresponsibility environmental destruction. The Grassroots Recycling Network explains the concept:

Zero Waste:

- Redesigns the current, one-way industrial system into a circular system modeled on Nature's successful strategies.
- Challenges badly designed business systems that "use too many resources to make too few people more productive".
- Addresses, through job creation and civic participation, increasing wastage of human resources and erosion of democracy.

- Helps communities achieve a local economy that operates efficiently, sustains good jobs, and provides a measure of self-sufficiency.
- Aims to eliminate rather than manage waste.

Zero waste strategies rely on a comprehensive approach which takes into account waste reduction throughout the lifespan of a product, from raw material to assembly line to re-use. High quality recycled materials can be substituted for mined or harvested resources. Waste reduction at the production stage includes more efficient use of feedstocks and changes in design which promote recycling. At the end of the cycle materials must be sorted, cleaned, and stockpiled for use in the same or similar products. Closing the production-recycling loop saves both the resource and the energy used in its extraction from the earth. Air pollution reductions of 20% to 95% are achieved when recycled feedstocks are substituted for raw materials. And the recycling of steel, aluminum, paper, and glass reduces water pollution and water use from 35% to 97%.

Burning lowers the quality of recycled feedstocks and diverts them to lower uses, breaking the recycling loop and increases the need for replacement resources from raw materials. For example, using incinerator ash or slag in cinderblocks locks up the materials which could have been re-used; incineration and gasification methods which do not separate aluminum, steel, and glass before processing aggravate this problem. The Global Recycling Council states that mixed garbage/recycling collection systems and high-temperature processing methods such as pyrolysis are incompatible with zero waste systems. Gary Liss, Secretary of the Council, said, “Certain conversion technologies represent an ‘end-of-pipe’ approach that could serve as yet another obstacle to waste reduction.” Gasification plans fail to reduce garbage volumes, divert money from recycling programs, and exacerbate the problems of waste.

Composting of solid waste utilizes a natural process to break down the organic fraction of household garbage. This low-tech method is inexpensive and produces a useful soil conditioner or mulch which returns organic matter to the earth. As other methods of waste management such as incineration fall out of favor, composting of municipal waste continues to grow nationwide. The United Nations Environmental Programme gives the following assessment:

> “Upwards of 5% of the MSW stream in North America is now managed through centralized composting programs, which were insignificant prior to the mid-1980s. The compostable portion of MSW can constitute 30-60% of a community’s waste stream. Composting programs have been designed for a variety of organic waste streams, including yard wastes (grass trimmings, leaves, or tree prunings), food wastes, agricultural wastes, and wastewater treatment sludge. Another alternative, which has been used only on a limited basis in North America, is mixed waste composting. Mixed waste processing facilities accept unsorted MSW in the same form as it would be received at a landfill or a waste-to-energy facility, and separate recyclable materials. The relatively small community of Guelph, Ontario, has been operating such a facility successfully since early 1996”.

Today, there are many municipal composting programs in the United States. Some are smaller operations handling less than 50 tons of waste per day. In 1995, rural Madison County, NC (population 17,000) initiated a composting project to demonstrate its practicability. Food wastes from school cafeterias, grocery stores, and restaurants, and old newspapers were composted, helping to reduce volume, toxic leachate, methane generation, and odor caused by 48 tons of solid waste per day.

Composting isincompatible with gasification because both require the same organic materials to operate. The compounds of carbon and hydrogen which comprise compostable waste would go up the smokestack of a gasification unit.

## Energy Balance of Electric Generation versus Recycling

Energy is wasted by gasification. Energy recovery in gasification is limited to the burning of a low-Btu gas which has about 25% of the energy value of natural gas, meaning four times as much synthetic gas must be burned to equal the heat of natural gas. Efficiency losses inherent in combustion reduces the recovered energy by 60-70%.

Energy recovery from waste burners of all types pales in comparison to the energy needed to manufacture new products. Paper made from trees requires double the energy of recycled paper. Each ton of recycled paper saves about two dozen trees and 410 gallons of fuel needed to produce new paper. And beverage cans made from aluminum ore require 20 times as much energy to produce compared cans made of recycled aluminum. Table lists the relative benefits of using recycled materials instead of virgin raw materials:

Table: Energy savings and environmental benefits of recycling.

| Reduction of: | Aluminum | Steel | Paper | Glass |
|---|---|---|---|---|
| Energy Use | 95% | 60% | 50% | 20% |
| Air Pollution | 95% | 85% | 74% | 20% |
| Water Pollution | 97% | 76% | 35% | - |
| Water Use | - | 40% | 58% | 50% |

According to the US Energy Information Administration, the aluminum industry's demand for electricity was 72,972 million kilowatt-hours (KWh) in 1998. A gasification cost analysis done by the Environmental Engineering Corporation states that a gasification unit using 72 tons of solid waste per day, or 23,652 tons per year, could produce excess power of 18,138 MWh per year, or 18.1 million KWh/year. ( 1 Megawatt-hour = 1000 Kilowatt-hours ) The EEC's analysis for "energy recovery" from solid waste would be 0.02 % of annual electricity usage for the aluminum industry. In other words, it would take 5,000 such gasification units to provide the power needed by the aluminum industry. the aluminum industry could save 69,323 million KWh/year (95% of 72,972 million KWh), 3800 times as much electric power, if all the aluminum was recycled. Unlike 5,000 new gasification units, 100% aluminum recycling is economical, practical and environmentally sound. Also in 1998, steel mills used 52,591million KWh, glass plants used 12,298 million KWh, paper mills used 50,241million KWh, and newsprint mills used 18,469 million KWh. Similar analyses with similar results can be shown for these industries.

Plasma, referred to as the "fourth state of matter," is a very high temperature, highly ionized (electrically charged) gas capable of conducting electrical current. Examples of plasma in nature include lightning and gas at the surface of the sun. Plasma technology has a long history of development and has evolved into a valuable tool for engineers and scientists who need to use very high temperatures for new process applications. Man-made plasma is formed by passing an electrical discharge through a gas such as air or oxygen ($O_2$). The interaction of the gas with the electric arc dissociates the gas into electrons and ions, and causes its temperature to increase significantly, often exceeding 6,000 °C, nearly as hot as the sun's surface.

Plasma gasification is an emerging technology which can process landfill waste to extract commodity recyclables and convert carbon-based materials into fuels. It can form an integral component in a system to achieve zero-waste and produce renewable fuels, whilst caring for the environment. Plasma arc processing has been used for years to treat hazardous waste, such as incinerator ash and chemical weapons, and convert them into non-hazardous slag.

Utilizing this technology to convert municipal solid waste (MSW) to energy is still young, but it has great potential to operate more efficiently than other pyrolysis and combustion systems due to its high temperature, heat density, and nearly complete conversion of carbon-based materials to syngas, and non-organics to slag.

Syngas is a simple fuel gas comprised of carbon monoxide and hydrogen that can be combusted directly or refined into higher-grade fuels and chemicals. Slag is a glass-like substance which is the cooled remains of the melted waste; it is tightly bound, safe and suitable for use as a construction material.

Plasma torch technology has proven reliable at destroying hazardous waste and can help transform environmental liabilities into renewable energy assets.

## Plasma Gasification Process

Plasma gasification is a multi-stage process which starts with feed inputs ranging from waste to coal to plant matter, and can include hazardous wastes. The first step is to process the feed stock to make it uniform and dry, and have the valuable recyclables sorted out. The second step is gasification, where extreme heat from the plasma torches is applied inside a sealed, air-controlled reactor. During gasification, carbon-based materials break down into gases and the inorganic materials melt into liquid slag which is poured off and cooled. The heat causes hazards and poisons to be completely destroyed. The third stage is gas clean-up and heat recovery, where the gases are scrubbed of impurities to form clean fuel, and heat exchangers recycle the heat back into the system as steam. The final stage is fuel production the output can range from electricity to a variety of fuels as well as chemicals, hydrogen and polymers.

Gasification has a long history in industry where it has been used to refine coal and biomass into a variety of liquid fuels, gases and chemicals. Modern clean coal plants are all gasifiers, and so were the earliest 19th century municipal light and power systems.

Plasma gasification refers to the use of plasma torches as the heat source, as opposed to conventional fires and furnaces. Plasma torches have the advantage of being one of the most intense heat sources available while being relatively simple to operate.

Plasma is a superheated column of electrically conductive gas. In nature, plasma is found in lightning and on the surface of the sun. Plasma torches burn at temperatures approaching 5500°C (10,000F) and can reliably destroy any materials found on earth with the exception of nuclear waste.

Plasma torches are used in foundries to melt and cut metals. When utilized for waste treatment, plasma torches are very efficient at causing organic and carbonaceous materials to vaporize into gas. Non-organic materials are melted and cool into a vitrified glass.

Waste gasification typically operates at temperatures of 150 °C (270 °F), and at those temperatures materials are subject to a process called molecular disassociation, meaning their molecular bonds are broken down and in the process all toxins and organic poisons are destroyed. Plasma torches have been used for many years to destroy chemical weapons and toxic wastes, like printed circuit boards (PCBs) and asbestos, but it is only recently that these processes have been optimized for energy capture and fuel production.

America's Westinghouse Corporation began building plasma torches with NASA for the Apollo Space Program in the 1960s to test the heat shields for spacecraft at 550 °C. In the late 1990s, the first pilot-scale plasma gasification projects were built in Japan to convert MSW, sewage sludge, and auto-shredder residue to energy. The Japanese pilot plants have been successful, and commercial-scale projects are under development now in Canada and other countries, by companies such as Alter NRG, from Alberta, Canada.

## Economics

The economics of MSW plasma gasification are favourable, although complex. Waste gasification facilities get paid for their intake of waste, via tipping fees. The system then earns revenues from the sale of power produced. Electricity is the primary product today, but liquid fuels, hydrogen, and synthetic natural gas are all possibilities for the future.

Sorting the MSW to capture commodity recyclables, such as metals and high-value plastics, presents a third revenue stream. Minor revenue streams include the sales of slag and sulphur. Slag has the potential to be used for a number of construction products, such as rock wool, bricks and architectural tiles, and sulphur has some commodity value as fertilizer.

Additional costs are avoided by diverting waste from landfills and minimizing transportation of waste. Government subsidies for renewable energy or carbon credits may be substantial in the future, but are difficult to project.

A base case scenario with a 680 tonne per day (750 US tons) waste gasification plant which would be appropriate for a small city or regional facility, would cost an estimated $150 million (€108 million) to construct. A municipality that funds the entire project through bonds should seek a positive cash flow year-after-year via revenues from tipping fees, recyclables and electricity sales, as well as sales of slag and sulphur. There is considerable range in the values for each of these variables, and any proposed development would require extensive due diligence to determine local prices for each line item. Tipping fees, electricity rates, commodity recyclables, as well as interest rates and taxes, all vary dramatically creating a model which needs to be thoroughly evaluated for any proposed development.

The economics of waste gasification heavily favour recycling inorganic materials like metal and glass have no value as fuel and make the gasification process less efficient, even though plasma torches have the ability to melt them. High-value plastics and papers that can be readily separated are far more valuable as recyclables than as fuel. Certain plastics earn €195 per tonne ($300 per US ton) and certain types of paper can earn around €53 per tonne ($75 per US ton). For comparison, a tonne of waste may produce 0.8 MW of electricity, worth around €51 ($70) per MW. It is clear that any of these materials that can be separated and sold, are worth much more as commodities than as fuel.

## Wide Variety of Inputs and Outputs

There are additional waste streams available in certain locations which earn higher tipping fees than MSW because they are toxic and yet have excellent fuel value. Refinery wastes from petroleum and chemical plants, medical waste, auto-shredder residue, construction debris, tyres and telegraph poles, are all examples of potential fuels that can earn high tipping fees and provide good heat value. Additionally, there are millions of tonnes of low-grade waste coal that exist in massive piles throughout the Appalachian region of Pennsylvania and West Virginia, US, that can be utilized for gasification.

Multiple outputs can be produced from a single facility. Heat and steam can be sold, and electricity production can be combined with ethanol or hydrogen production to maximize resources. Hydrogen can be readily produced from syngas by separating it from the carbon and oxygen, while synthetic natural gas can be produced by upgrading the methane content of syngas.

Liquid fuels are typically produced from syngas through catalytic conversion processes such as Fischer-Tropsch which has been widely used since World War II to produce motor fuels from coal. Biotech methods to produce liquid fuels are also being developed to use enzymes or micro-organisms to make the conversion.

Much research and effort is being put into developing more selective catalysts and productive enzymes which will raise system efficiencies to levels needed to be competitive. Currently, ethanol from gasification costs more than $2 a gallon (equivalent of €0.37 per litre), and it is estimated that production needs to cost closer to $1.25 (€0.90) or $1.50 (€1.10). Production of ethanol at demonstration scale has shown that one US ton of MSW can produce around 100 gallons (equivalent of 0.9 tonnes producing 380 litres) of ethanol, give or take 20%. Cost estimation for ethanol production is difficult, but rough calculations indicate that ethanol could potentially be more profitable than electricity.

## Improved Waste Management

Gasification is superior to landfilling MSW for a number of reasons. First of all, landfills are toxic to the environment due to the production of toxic liquid leachate and methane gases. The EPA (US Environmental Protection Agency) has a lengthy protocol for airborne and liquid chemicals which must be contained and monitored for every landfill. Landfills must be constructed with extensive liners, drains and monitoring equipment to comply with regulations. Plasma gasification can divert waste from landfills and create beneficial uses for the material, by maximizing recycling and cleanly using the rest for fuel.

## Gasification is Superior to Incineration

Gasification is superior to incineration and offers a dramatic improvement in environmental impact and energy performance. Incinerators are high-temperature burners that use the heat generated from the fire to run a boiler and steam turbine in order to produce electricity. During combustion, complex chemical reactions take place that bind oxygen to molecules and form pollutants, such as nitrous oxides and dioxins. These pollutants pass through the smokestack unless exhaust scrubbers are put in place to clean the gases.

Gasification by contrast is a low-oxygen process, and fewer oxides are formed. The scrubbers for gasification are placed in line and are critical to the formation of clean gas, regardless of the regulatory environment. For combustion systems, the smokestack scrubbers offer no operational benefit and are put in place primarily to meet legal requirements. Plasma gasification systems employing proper scrubbers have extremely low emissions and no trouble meeting and beating the most stringent emissions targets.

The objective of gasification systems is to produce a clean gas used for downstream processes which requires specific chemistry, free of acids and particulates so the scrubbing is an integral component to the system engineering, as opposed to a legal requirement that must be met.

Incinerator ash is also highly toxic and is generally disposed of in landfills, while the slag from plasma gasification is safe because it is melted and reforms in a tightly-bound molecular structure.

In fact, one of the main uses for plasma torches in the hazardous waste destruction industry has been to melt toxic incinerator ash into safe slag. The glassy slag is subject to EPA Toxicity Characteristic Leaching Procedure (TCLP) regulations that measure eight harmful elements. Data from existing facilities, even those processing highly hazardous waste, has shown them to be well below regulatory limits.

Electricity production from plasma gasification is superior to that from incinerator combustion. Incinerators typically use the heat from combustion to power a steam turbine to produce power. Gasification systems can use gas turbines that are far more efficient, particularly when configured in integrated gasification combined cycle mode (IGCC). Just as IGCC is the state-of-the-art in producing power from coal, the same is true when using MSW as the fuel source.

## Carbon Impact

The carbon impact of plasma gasification is significantly lower than other waste treatment methods. It is rated to have a negative carbon impact, especially when compared to allowing methane to form in landfills. Gasification is also an important enabling technology for carbon separation. It is primarily a carbon processing technology; it transforms solid carbon into gas form.

Syngas is comprised of carbon monoxide and hydrogen. The hydrogen readily separates from the carbon monoxide allowing the hydrogen to be used while the carbon is sequestered. The US Department of Energy has identified gasification through its clean coal projects as a critical tool to enable carbon capture.

## Environmental Opposition

Environmentalists have expressed opposition to waste gasification for two main reasons. The first argument is that any waste-to-energy facility will discourage recycling and divert resources from efforts to reduce, reuse and recycle. Economic studies of the waste markets show the opposite to be true; waste-to-energy heavily favours the processing of waste to separate valuable commodities and to maximize its value for fuel.

The second argument made against waste gasification is that has the same emissions as incineration. These arguments are based on gasification systems which do not clean the gases and instead combust dirty syngas. Such systems are essentially two-stage burners and are not recommended for environmental reasons. There are many variations of combustion, pyrolysis and gasification all used in different combinations. Proper engineering is required to achieve positive environmental performance.

## Efficiency of Plasma Gasification Technologies

It is well known that one of the most effective hazardous waste destruction technologies is a thermal treatment and gasification. However, there exist an entire group of substances any traditional treatment of which causes a threat to the environment and human health. Therefore, present research proposes to develop and implement plasma technology, which allows to remove all waste containing hazardous substances. Plasma decontamination technology of toxic materials allows to create a compact device, which can reliably neutralize all of types of hazardous waste. Such plasma device is characterized by a very high temperature, short reaction time, extreme activation energy, the ability to heat various gases, effective neutralization and independence from fuel sources.

Complete and safe hazardous waste (outdated medications, banned pesticides, plastic gears, pathological waste, container, etc.) removal method is high temperature (plasma) pyrolysis. It is already employed in many countries around the world: USA, Japan, France, Germany, Switzerland, Australia, etc. Many developing countries (India, China, Belarus, etc.) also seek to employ the plasma technology in this area. There is a shortage of detailed technology description in worldwide scientific literature because these research results some times are not made public. Medical wastes from hospitals, dental clinics and other health centers are collected and recycled in about 1500 large companies. Most of them are located in the USA, France, Great Britain and Japan.

The interest in plasma technology in the application of harmful substances neutralization processes is huge. For example, the Japanese medical waste management company recently implemented a large project, whose main goal is to transform the infected local medical waste into useful products - glass, metal and syngas.

The most important problem is their energy efficiency, because the efficiency of electricity generation to power the plasma torch (PT) is only about 30%. Thus, in order to achieve the commercialization of such environmentally clean technologies, they need to simultaneously achieve high levels of their energy efficiency.

## Waste-to-energy Process

Modern technologies of the waste treatment are oriented on the processes of their gasification. It has three interrelated advantages. First of all, the temperature range at which the gasification processes are effectively carried out is quite high and usually exceeds 1000 °C. This automatically meets the requirements according to which the temperature should be maintained at 1100 °C in case of incineration of waste containing more than 1% wt. of halogenated organic substances under conditions of chloride. This is necessary for dioxins and furans which are formed at lower temperatures, to be effectively decomposed into HCl.

Second, each local volume of gas produced in the processing has to be kept at this temperature over time ≥2 s. In this case, maximum permissible emission of dioxins and furans to the atmosphere in the refinement products do not exceed $10^{-10}$ g/m$^3$. This is very important as these compounds are among the most toxic ones. In addition, prolonged residence of reagents at high temperature ensures the completeness of gasification processes, and also allows accepting the assumption of equilibrium conditions when performing thermodynamic calculations.

Third, although gasification products must be cooled down quickly to avoid the reverse generation of dioxins and furans, the main energy is accumulated in chemical bonds. Even though syngas cooling leads to some losses of thermal energy, the share of which is small compared to the total energy content which consists of thermal and chemical energies.

Another problem appears when the waste contains in its composition heavy metals; using well known incineration for their utilization leads to formation of ash, which is itself a hazardous waste. The latter environmental hazard is particularly dangerous in the case of recycling the sewage sludge of urban wastewater treatment plants.

The arc discharged plasma is an effective tool for many types of application including hazardous waste treatment. It is important to notice that there exist several unsolved problems in thermal treatment of sewage sludge area. During the combustion process, solid dispersed particles may be formed from the combustion products. Solid particles may penetrate into the human lungs and can cause serious illnesses. Incomplete combustion may also occur inside the furnace and form new chemicals that may appear to be more toxic than initial material. Therefore, flue gas is cleaned in multicyclones or fiber filters before discharge into the atmosphere. However, such types of filters are expensive and not very effective in the case of fine dispersed particulates. There does not exist means against newly formed hazardous chemicals at all. So the plasma treatment of exhaust combustion products is welcomed. Atmospheric pressure arc plasma is also a promising tool for the synthesis of catalytic coatings which could be successfully employed in the manufacturing of catalyst for flue gas treatment.

## Plasma Processing of Hazardous Waste

First in Ukraine, full-scale equipment for medical waste processing as well as another hazardous waste has been built by the E.O. Paton Electric Welding Institute of the National Academy of Sciences of Ukraine (NASU) and the Institute of Gas, NASU. Its fundamental advantage is using water steam-plasma as a gasification agent, which allows to obtain the gasification products of maximum calorific value.

The other type of experimental equipment for destruction of hazardous waste has been installed in Lithuanian Energy Institute (LEI). It consists of a plasma jet reactor with DC arc plasma source capacity of up to 90 kW. The plasma process uses air, nitrogen, water vapor or their mixtures. The plasma-forming gas flow rate in the reactor reaches up to 2–7 g/s, the average exhaust mass temperature varies from 2800 to 3500 K. Experimental and numerical studies carried out upon the realization of the plasma decomposition process of organic and inorganic substances.

## Plasma Sources

Arc plasma torch (PT) is a key element of the equipment. It was made according to the two-electrode axial scheme with hollow copper electrodes. Compressed air and steam are used as the plasma-forming gases. PT ignition is carried out with air and then transition to steam occurs after the heating.

The linear DC arc heater was produced in LEI for heating air, nitrogen, steam or their mixtures up to 7000 K. It was connected to the reactor vessel. By achieving gas temperature over 4000 K, molecules of hazardous substances and waste decay to atoms, radicals, electrons and ions so that it appears ability to obtain simple combination of harmless chemicals. Several configurations of linear DC PT with hot cathode and step-formed anode were considered. As a sample, it could be mentioned PT 70 kW of power, with radial and tangential injection designed especially for the production of non-equilibrium plasma jet. The novel PT was manufactured and applied for the treatment of hazardous organic and inorganic compounds. It consists of a button type hafnium cathode, transitional copper anode for arc initiation 3, neutrode 5, insulation rings and step-formed copper anode 7. To increase the angular velocity of arc rotation, magnetic stabilization of flow was applied employing the coil 8.

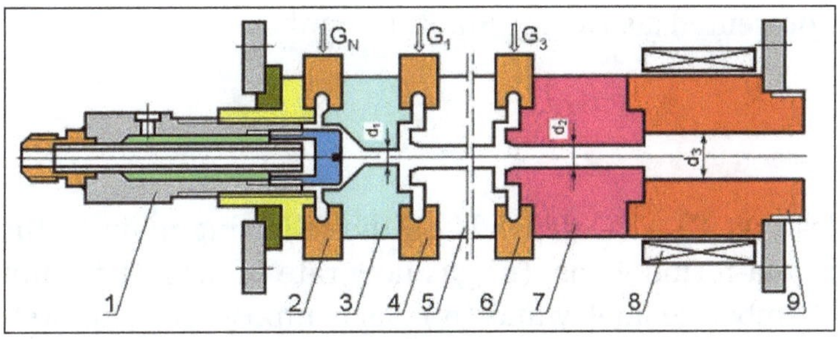


Schematic presentation of linear DC plasma torch.
1—Cathode junction with hafnium emitter; 2, 4, 6—Insulating rings with gas injection;
3—Intermediate anode; 5—Neutrode; 7, 9—Step-formed anode; 8—Magnetic coil.

A new PT is employed for heating the material that is injected into the reaction arc chamber. Both average and local heat losses of PT elements are necessary to know when the device is operating under extreme conditions to select operating and cooling regimes. Operating characteristics of the PT plasma flow and parameters were determined from the heat conservation calculations while measuring voltage drop, gas flow rate and arc current strength in the circuit. The preference has been given to the PT with neutral, fixed average arc length and step-formed copper electrode. This enabled to reduce arc shunting after anode step and ensured the stability of length of the arc in the wide diapason of gas flow rates and current variation. The employed plasma source also is different comparing to ordinary plasma torches with the conical expanded anode. The anode step also serves for reduction of the pressure drop in the discharge channel and to fix the arc in the stable position. The total PT length is 0.25 m, the insular part anode diameter is 0.03 m and the diameter of extended part of the anode is 0.04 m. The neutrode makes separate neutral section of the torch and is isolated from the anode. It is located between insulating rings made of thermal resistant glass textolite. Each ring is also used for tangential air supply and contains a pair of tangential-oriented blowholes for the arc stabilization. The experimental equipment for producing arc plasma is comprised of rectifier for power supply, gas supply, water-cooling systems and airing devices.

The modified similarity theory has been applied for the analysis of operating and thermal characteristics and result generalization. Voltage–current characteristic (VCC) of PT were generalized employing criterial equations and following expressions were established:

$$\frac{Ud_2}{I} = 1350\left(\frac{I^2}{Gd_2}\right)^{-0.55}\left(\frac{G}{d_2}\right)^{-0.14}(pd_2)^0.$$

PG performance and thermal characteristics can be evaluated by its efficiency η indicating what part of generated energy is transferred to gas:

$$\eta = \mathrm{GH(UI)}.$$

Generalization of the TC of PG is similar to generalization of the electric characteristic:

$$\frac{1-\eta}{\eta} = 5.5\times10^{-3}\left(\frac{I^2}{Gd_2}\right)^{0.22}\left(\frac{G}{d_2}\right)^{-0.12}(pd_2)^0\left(\frac{l}{d}\right)^0.$$

Here U is arc voltage, I is arc current, G is total gas flow rate, $d_2$ is anode diameter and p is pressure. The value of η may be presented also as the Stanton number:

$$\frac{1-\eta}{\eta} = \frac{4l}{d_2}St.$$

The research concludes that PT VCC depends on the following main factors: (i) radial and tangential injection of plasma-forming gas; (ii) gas flow rate of plasma-forming gas to produce the desired arc; (iii) arc chamber geometry and (iv) gas composition. The first factor was evaluated during the experimental investigation of gas flow rate at the constant and various values of PT. In the present and previous studies when the radial injection is not applied, operating characteristics were observed as decreasing in the current range between 150 and 250 A. This follows as a result of dropping electric field intensity which linearly depends on the arc current. It was also established that voltage drop and electric field intensity linearly decrease with increasing of gas flow rate in the range of 7–10 × $10^{-3}$ and 5–8 × $10^{-3}$ kg $s^{-1}$. When the radial and tangential injection in different locations is used, the arc is strongly turbulized and a possibility to heat up much larger amount of gas in the PT of reduced dimensions is available. Consequently, the voltage drop in such PT increases up to 70% and the possibility for better control of plasma-forming process appears.

When tangential injection of plasma-forming gas is applied inside the PT anode, the character of operating characteristics is slightly dropping or remains as stabile. It is important to notice that static PT characteristics may be also slightly rising with increase of arc current strength.

The present measurements over 120 experiments were carried out varying with the help of resistors arc current strength and injected air flow rate *G1* and *G3*.

## Plasma Chemical Reactors

Technologically, the conversion process is carried out in a flow reactor. It has a metal case and is lined with the layer of fireproof and heat-insulating materials on the inside. PT electrical power reaches up to 160 kW with efficiency coefficient = 0.7–0.8. The equipment also includes lock-chamber for the periodic load of the packed medical waste, steam generator, power supply of up to 500 volts and a current up to 350 A, as well as the system for the gas quenching and cleaning.

Table: Plasma source technical parameters.

| Power, P (kW) | 33–78 |
|---|---|
| Arc current, U (A) | 175–245 |
| Arc voltage, I | 160–335 |
| Cooling water flow rate, Gv (kg $s^{-1}$) | 0.16–0.18 |
| Water temperature increment (deg): | |
| plasma torch | 15–23 |
| cathode | 1.1–1.53 |
| ignition section | 1.08–2.16 |
| neutrode | – |
| anode | 13.0–19.3 |
| Source gas flow rate (kg $s^{-1}$): | |
| cathode, $G_N$ | 0.54–1.0 |
| neutrode, $G_1$ | – |
| anode, $G_2$ | 1.85–7.6 |
| Plasma jet average mass temperature (K) | 3460–5200 |

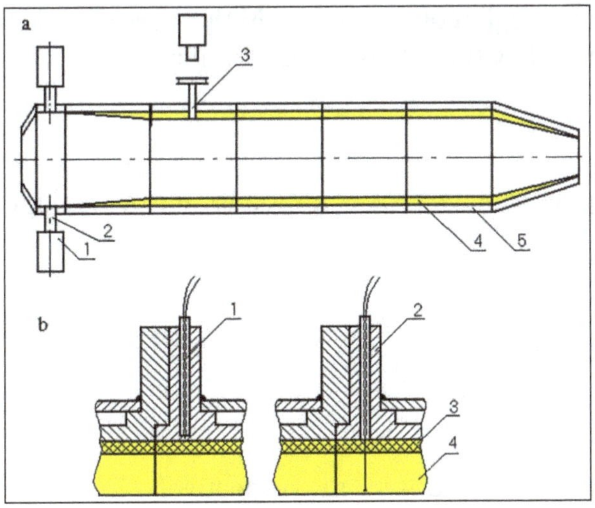


Schematic presentation of plasma jet reactor for treatment hazardous waste. (a) Stream reactor with: 1—Plasma torch; 2—Plasma torch and feeder connecting section; 3—Window for observation and measurement; 4—Layer of $Zr_2O_3$; 5—Cooling section (five units). (b) Construction of thermocouple's junction: 1—Thermocouple; 2—Frame; 3—Layer of insulating cover; 4—layer of ceramic cover.

The lock-chamber for medical wasteload is located in the upper part of the reactor. Unit management does not involve the full loading of the total reactor space with wastes. This is important for gasification products, if they move through a thick layer of raw materials, not to be cooled below 1100 °C.

Table presents the composition of the basic gasification products obtained from the medical waste in the equipment for plasma-steam gasification. In these experiments, organic wastes of such average composition have been studied: 60% of cellulose $C_6H_{10}O_5$ + 30% of plastics based on polyethylene $(-CH_2-CH_2-)n$ + 10% water.

The main physical result of this experimental exploration was a possibility of self-power supply by syngas with gas-diesel engine system taking into account even low efficiency of electricity production ~30%. In general, the previous experience of using this equipment has confirmed the correctness of the basic technical solutions laid down therein. However, it also revealed some shortcomings of individual design solutions. They demand the revision process of further development. In particular, this applies to the high temperature thermal insulation of the reactor.

Three different plasma chemical reactors were designed in LEI:

- Straight stream reactor for flue gas treatment.
- Curved stream reactor for the treatment of gaseous, liquid and solid substances with small solid dispersed particles.
- Steady ARC volume reactor, devoted for incineration of wide range of waste.

Assuming that the plasma flow has been characterized as optically thin. The transport coefficients and thermodynamic properties depend only on the temperature and pressure. The plasma flow in the reactor is also characterized with extremely high temperature gradients and recirculating turbulent flow with wall confinement. The flow inside the chamber was separated. Heat transfer characteristics in the entrance region of the reactor in this case of sudden expansion for the region of $x/d < 0.4$ could be described by the following equations:

$$Nu_{fd} = 0.006\,\mathrm{Re}_{fd}^{0.86}.$$

For the region of x/d > 0.4 described by the equation for entrance region of the pipe:

$$Nu_{fd} = 0.0256\,\mathrm{Re}_{fd}^{0.8}\,\varepsilon l.$$

Here εl is the entrance factor, equal:

$$\varepsilon l = 1.48(x/d)^{-0.15}.$$

Nu and Re are Nusselt and Reynolds criterions, respectively. Index fd means that Nu and Re are calculated according to the flow conditions in the entrance and reactor channel diameter.

## Plasma Application in Sewage Sludge Treatment

During sewage treatment, the main pollutants are separated as sewage sludge. Depending on the original pollution load of the water being treated, they may include the heavy metals in their composition. The Kyiv wastewater treatment plant (known as Bortnychi station of aeration) processes municipal and industrial sewage and run-off rain water. It accepts 9000 $m^3$ wastewater per day on an average. At present, 9 million tone of sewage sludge are accumulated on its territory.

Centralized wastewater treatment plants in Lithuania produce relatively small amounts of sewage sludge. The annual amount of dry sewage sludge produced in Lithuania is up to 50 thousand tons per year.

The special problem of this waste is heavy metals in its compound. The presence of these pollutants prevents the burial of sewage sludge and substantially limits its use in agriculture and forestry. A similar situation occurs when certain wastes (e.g., industrial, medical, military and sewage sludge) are destroyed in special devices known as incinerators, which leads to the formation of relatively high toxic waste in ash. Toxic residues (ash, slag, sediment of filters and sedimentation tanks) can be easily placed on landfills in case they were first immobilized and converted to non-leachable products. If these residues are heated to a very high temperature, then their main components, including minerals and toxic heavy metals, melt and take on a glassy appearance. This requires temperatures above 1700 K, which are not available in the most incinerators, but are easily achieved in plasma reactors. The system of plasma vitrification of ash produces a chemically stable and mechanically strong substrate. After vitrification, this mineral product looks like a vitreous, similar in structure to basalt lava (even superior to basalt by mechanical strength); its main components are oxides of silicon, aluminum and calcium in the form of chemically inactive compounds that are resistant to washing. The effectiveness of this technology is convincingly confirmed by the data on the example of vitrification of the ash residue in a medical incinerator.

A simple empirical estimate of the energy inputs required for the vitrification process is:

$$\mathrm{M(kg)} = 0.35\mathrm{P(kWh)},$$

where $M$ is the mass of the vitrified product and P is the electrical energy consumed in the process. It is quite simple and allows you to calculate the energy required for the gasifier, regardless of the thermodynamic calculations associated with the conversion of carbon-containing raw materials.

Thermodynamic calculation of the gasification process using plasma technologies.

## Generalized Reaction of Gasification

At present, quite a lot of software tools have been developed and used for quantitative analysis of gasification processes. However, with all the advantages of numerical calculations, such publications leave "in shadow" basic physical and chemical regularities. Just the knowledge of their characteristics built a clear understanding of the analyzed process. In reality, the basis of the quantitative description of gasification lie very simple thermodynamic relations arising from the laws of Hess used in thermochemistry. It should be borne in mind only the features associated with the operation of the plasma source.

The process of plasma-steam gasification can be represented by the gross equation in a sufficiently general form:

$$CH_xO_y + wH_2O + mO_2 = n_1H_2 + n_2CO + n_3CO_2 + n_4H_2O + n_5CH_4 + n_6C + Q_{TR},$$

where $Q_{TR}$ = QR + ΔQ is the total thermal energy that is released as a result of the chemical reactions QR and due to some additional source of heat energy ΔQ (so far we do not necessarily associate it with the energy of the plasma jet QPL), so that the reaction mixture reaches the desired temperature TP of the gasification products, w and m—the amount of water and oxygen, per 1 kmol of waste, $n_1$, $n_2$, $n_3$, $n_4$, $n_5$ and $n_6$ are the coefficients for the corresponding reaction products. Among the latter are gases, most often obtained in the composition of gasification products and soot. It allows to distinguish the role of an additional source of energy ΔQ in viewpoint of achieving the optimal, predictably perceived, temperature TP of the gasification process.

The "ideal" process of plasma-steam gasification would correspond to the case when only $H_2$ and CO would be present on the right side. Formally, it is possible to make many options of reaction $CH_xO_y + wH_2O + mO_2 = n_1H_2 + n_2CO + n_3CO_2 + n_4H_2O + n_5CH_4 + n_6C + Q_{TR}$, with various stoichiometric coefficients, including the relevant "ideal" process. However, in accordance with the second law of thermodynamics, nature chooses only such a path and the completion of the varying reactions, in which the principle of maximum entropy is realized:

$$dS \geq dQ / T.$$

Special software—"TERRA" thermodynamic calculations system is used for the conversion processes quantitative analysis with a glance of the accompanying reactions. It also allows to determine the necessary amount of energy expenditure for carrying out reactions.

## Plasma-steam Gasification

Analysis of the process of plasma-steam gasification was made on a more optimal than $CH_{2.483}O_{0.530} + 1.334\ H_2O = 2.549H_2 + 0.111CO + 0.876CO_2 + 0.013CH_4$. reaction:

$$CH_{2.5}O_{0.5} + 0.5H_2O = CO + 1.75H_2 + Q_{TR}.$$

The heat of combustion of sewage sludge $Q_{LSS}$ required to determine the energy of the process is determined on the basis of Mendeleev's equation:

$$Q_{ISS} = -100 \bullet (81 \bullet c_C + 246 \bullet c_H - 26 \bullet (c_O - c_S) - 6 \bullet c_W) \bullet 4.19 kJ / kg,$$

where $c_C$, $c_H$, $c_O$, $c_S$ and $c_W$ are mass fractions of carbon, hydrogen, oxygen, sulfur and water. (Mendeleev's equation is an analog of the relations known as Dulong or Milne equation.) As may be shown, this heat of combustion of sewage sludge is $Q_{LSS}$ = −25.68 MJ/kg. Following the law of Hess, its enthalpy of formation is $\Delta H^o_{CH2.5O0.5}$ = −76.8 kJ/kg.

The thermodynamic analysis of the sewage sludge conversion process carried out in the TERRA software allows to determine the composition of its gasification products as a function of temperature. As it turned out, both for the reaction $dS \geq dQ / T$. and for other considered reactions, it is characterized by the practical completion of the gasification processes at 1250 K. More strictly, the mass fraction of the traces of $CO_2$, $H_2O$ and $CH_4$ among the products of gasification at this temperature does not exceed 1–2%. As it turned out, the energy $Q_{PL}(T)$, which must be additionally introduced with a plasma torch per 1 kg of reagent mass in $dS \geq dQ / T$., to reach this temperature, is 0.785 kWh/kg. This parameter allows to determine the productivity of the gasifier at a given power of the PT.

Knowing the calorific values for CO and $H_2$, as well as the composition of the products obtained in the reaction $dS \geq dQ / T$., it is easy to determine the calorific value of the resulting syngas in this process $W_{SG}$ = 6.23 kWh/kg. It allows to define the energy output of the gasification plant and its energy efficiency on the basis of a comparison with the specific energy QPL introduced into the reactor.

The value WSG significantly exceeds the electricity consumption 0.785 kWh/kg by steam PT to produce 1 kg of syngas. Thus, even taking into account the relatively low efficiency of $\eta_{EE}$ ~ 0.3 of electricity generation, the energy consumption is much lower than the level of energy of syngas produced. Indeed, taking into account also the efficiency of the PT at $\eta_{PL}$ = 0.8, this is enough to ensure the operation of the PT, since it exceeds the value of ΔQ = 0.785 kWh/kg:

$$W_{SG} \bullet \eta_{EE} \bullet \eta_{PL} = 6.23 \times 0.3 \times 0.8 = 1.5 \text{kWh / kg} > 0.785 \text{kWh / kg}.$$

It is good preconditions for the energy self-sufficiency of the sewage sludge processing and the production of additional energy to compensate the role of raw materials moisture and ash residue vitrification or for the production of electricity for external consumers.

## Plasma-steam-oxygen Gasification in Stoichiometric Mode

Significant increase of conversion efficiency can be achieved by the addition of oxygen into the process. At the first stage, an "ideal" conversion reaction was considered, in which the number of reagents, in contrast to $CH_{2.5}O_{0.5} + 0.5H_2O = CO + 1.75H_2 + Q_{TR}$., contained also oxygen, and among the reaction products syngas components only were present:

$$CH_{2.5}O_{0.5} + KH_2O + LO_2 \rightarrow CO + MH_2 + Q_{TR}$$

where K, L, M are coefficients that determine the content of components such as steam and oxygen, as well as the hydrogen one in the reaction products, respectively, under the stoichiometric reaction with respect to syngas production. Thus this reaction is stoichiometric as well as $CH_{2.5}O_{0.5} + 0.5H_2O = CO + 1.75H_2 + Q_{TR}$. for obtaining products of gasification as synthesis gas only. Nevertheless it has the most wide functional possibilities to achieve the best index of energy efficiency of the process as it allows varying the composition of the gasification agent. In determining the energy efficiency, naturally, the consumption of energy for oxygen production should also be taken into account. The range of possible specific energy consumption in the technological process of obtaining the oxygen itself is chosen as $PO_2$ = 0.35–1 kWh/m³. The first one corresponds to promising technologies, the second one is realistic today. Quantitative index of energy efficiency of the conversion process is the ratio:

$$\eta = \left(P_{PL}^{C} + PO_2\right) / W_{SG},$$

Where $P_{PL}^{C} = \Delta Q/0.8$ is the electricity consumption for the production of plasma jet by efficiency of $\sim 0.8$ and for oxygen – $PO_2$. WSG is the heat energy of syngas from 1 kg of the original raw mixture. In this form, it fully corresponds to the definition of energy saving (or energy efficiency) as energy costs (here, *PPLJ* + $PO_2$) per unit productivity (here the product is syngas of energy $W_{SG}$).

The value of $L$ = 0 in reaction $CH_{2.5}O_{0.5} + KH_2O + LO_2 \rightarrow CO + MH_2 + Q_{TR}$ corresponds to the plasma-steam gasification $CH_{2.5}O_{0.5} + 0.5H_2O = CO + 1.75H_2 + Q_{TR}$., and the case $\Delta Q$ = 0 is usual steam-oxygen technology, although their opposition does not make sense. Indeed, from the point of view of the process chemistry, in both cases, oxygen atoms, characteristic of these technologies, and hydrogen atoms, originally included in the gasified sewage sludge, are present in the reaction. For the noted limit values of *L*, the coefficient *K* takes the values $K_{max}$ = 0.5 and $K_{min}$ = 0, respectively. However, generally speaking, the reactions $CH_{2.5}O_{0.5} + KH_2O + LO_2 \rightarrow CO + MH_2 + Q_{TR}$ can also correspond to the intermediate values of the coefficients *K* and *L*. Simple functions are determined on the basis of mass balances in reaction $CH_{2.5}O_{0.5} + KH_2O + LO_2 \rightarrow CO + MH_2 + Q_{TR}$:

for oxygen:

$$1 \bullet 0.5 + K + 2L = 1, \text{or } L = 0.25 - 0.5K;$$

for hydrogen:

$$1 \bullet 2.5 + 2K = 2M, \text{or } M = 1.25 + K.$$

For clarity, the function $1 \bullet 0.5 + K + 2L = 1, \text{or } L = 0.25 - 0.5K;$, which characterizes the oxygen content of *L* as a function of the amount of steam *K* introduced by the PT. Line 4 represents the thermal power introduced into the reactor by a plasma jet at its nominal enthalpy of $H_{PL}$ = 3.6 kW·h/kg in accordance with equation.

$$Q_{PL} = H_{PL} mH_2O,$$

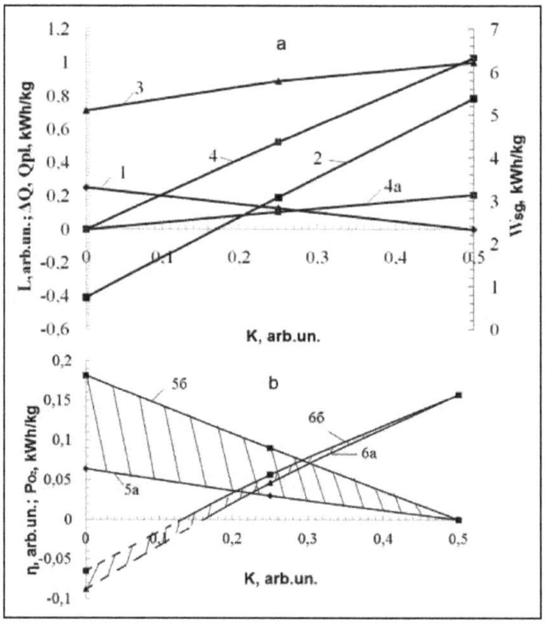


The main regularities characterizing the stoichiometric mode of gasification of the sewage sludge in the function of the amount of water introduced into the reaction K with plasma-steam jet – molar and energy ratios (a) and energy consumption for oxygen production and energy efficiency indicators of process (b): 1—oxygen content L introduced into the reactor; 2—additional energy ΔQ, which should be introduced in volume to achieve the operating temperature; 3—the energy of the producing syngas WSG; 4, 4a—the energy introduced by the steam-plasma jet QPL with its enthalpy HPL = 3.6 and 0.72 kWh/kg, respectively; 5a and 5b—energy consumption for oxygen production at a specific consumption of energy 0.35 and 1 kWh/Nm³, respectively; 6a and

6b—indicators of energy efficiency of the process at the corresponding specified energy consumption for the production of oxygen.

Where $mH_2O$ corresponds to the mass of water in the jet injected per kg of reagents. This enthalpy value corresponds to the moderate operating mode of the PT. In principle, the higher values of plasma enthalpy, corresponding to the forced operating regime of the steam PT, can be achieved.

It can be concluded that the introduction of oxygen in the stoichiometric mode of gasification with the use of plasma technologies corresponds to an increase in the energy efficiency of the process. As it follows from figure, the maximum value η in the process (which corresponds to the highest value of the additional energy ΔQ and, consequently, the worst energy efficiency) occurs exactly when the oxygen content is L = 0, for which K = 0.5 corresponds—that is, on the right side of each graph. On the contrary, the value η decreases with a gradual increase of the oxygen content L (i.e., moving to the left along the abscissa).

Area K < 0.17 corresponds the negative values ΔQ < 0; this means that excess heat energy is released in the reaction zone, which can be used for the ash residue's vitrification. The level of energy consumption for this need is difficult to determine in general terms, but empirical ratio $CH_{2.5}O_{0.5} + 0.5H_2O = CO + 1.75H_2 + Q_{TR}$. is known for them. In this area, lines 6 characterizing the level of energy efficiency of the process η are indicated by a dashed line. This is emphasized by the fact that here the energy costs for maintaining the gasification process are negative. In other words, there is an energy release.

In the absence of a PT, the stoichiometric gasification regime according to the reaction $W_{SG} \bullet \eta_{EE} \bullet \eta_{PL} = 6.23 \times 0.3 \times 0.8 = 1.5 kWh/kg > 0.785 kWh/kg$ is realized for a single value Ko = 0.17, corresponding to the intersection of line 2, which characterizes the required energy level ΔQ with the coordinate axis. It, in turn, corresponds to the moisture content of sewage sludge of about 10%, if it is determined from the composition of the reagents on the left side of the reaction $W_{SG} \bullet \eta_{EE} \bullet \eta_{PL} = 6.23 \times 0.3 \times 0.8 = 1.5 kWh/kg > 0.785 kWh/kg$ This moisture value is characteristic just for the conditioning of sewage sludge, which are currently dried with the help of those or other drying technologies.

However, the range of values K = 0.17 and near it for practical operation of the gasifier should be excluded, because the software TERRA reveals a significant soot formation, which makes it unacceptable for gasification. Thus, solving also the problem of obtaining more high quality syngas, it is expedient to move along line 2 to its maximum value corresponding to the stoichiometric regime at K = 0.5. As can be concluded, an increase in the amount of water introduced into the process L is twice, corresponds to a worsening of energy efficiency of the conversion process η by a factor three.

Table: Calculated parameters characterizing the stoichiometric process of conversion of sewage sludge at its humidity of 10% to synthesis gas using plasma technology, depending on the additional amount of water vapor introduced with the plasma jet.

| Parameter | K, arb.un. | |
|---|---|---|
| | 0.25 | 0.5 |
| L, arb.un. | 0.125 | 0 |
| ΔQ, kWh/kg | 0.19 | 0.785 |

| | | | |
|---|---|---|---|
| $P_{PL}{}^{C}$, кВт·ч/кг | | 0.24 | 0.98 |
| $PO_2$, kWh/kg | $PO_2$ = 0.35 kWh/Nm³ | 0.03 | 0 |
| | $PO_2$ = 1 kWh/Nm³ | 0.09 | 0 |
| $W_{CT}$, kWh/kg | | 5.79 | 6.23 |
| η, arb.un. | $PO_2$ = 0.35 kWh/Nm³ | 0.046 | 0.15 |
| | $PO_2$ = 1 kWh/Nm³ | 0.057 | 0.16 |

## Non-stoichiometric Mode of Plasma-steam-oxygen Gasification

The introduction of a significant amount of energy with a plasma jet markedly worsens the indicator of the energy efficiency of the plant. Therefore, it is of interest to compare it with the non-stoichiometric regime, which can be easily realized for the same value of Lo = 0.165, as in the stoichiometric regime at the point Ko = 0.17, but at K > 0.17. Therefore, it is advisable to introduce excess oxygen into the reactor.

In this case, in order to optimize the plasma-steam gasification process of sewage sludge, the next reaction was analyzed:

$$CH_{2.5}O_{0.5} + KH_2O + LO_2 \rightarrow CO + MH_2 + ECO_2 + DH_2O + Q_{TR}.$$

This gasification mode is called "non-stoichiometric", as there are the products of partial combustion of sewage sludge—$CO_2$ and $H_2O$—among the products of gasification. In determining the parameters of the process in the non-stoichiometric mode of gasification, it should be taken into account that, in addition to the syngas, the ballast components are formed from the unit mass of the initial reagents. In other words, the correction factor should be taken:

$$W^{*}_{SG} = [(mCO + mH_2)/(mCO + mH_2 + mH_2O)]W_{SG} = k_{NS}\, W_{SG},$$

where kNS is the non-stoichiometric coefficient.

Recall that, in principle, it is possible to compose many variants of the reaction $CH_{2.5}O_{0.5} + KH_2O + LO_2 \rightarrow CO + MH_2 + ECO_2 + DH_2O + Q_{TR}$. with the different stoichiometric coefficients. However, in fact, only those are actually realized where maximum entropy principle is satisfied.

Table: Composition of gasification products in non-stoichiometric mode with oxygen content *Lo* = 0.165 according to the calculated data in the TERRA software.

| The water content in reagents, the mole fraction of K | Composition of gasification products, wt. | | | |
|---|---|---|---|---|
| | CO | $H_2$ | $CO_2$ | $H_2O$ |
| 0.25 | 0.84 | 0.089 | 0.034 | 0.032 |
| 0.5 | 0,.69 | 0.083 | 0.11 | 0.117 |

Using these data, the parameters of non-stoichiometric gasification regimes for K = 0.25 and 0.5 were calculated.

Analyzing the results presented, it should be borne in mind that they are not energetically self-consistent. Indeed, with an oxygen content Lo = 0.165, a relatively small additional thermal energy

$\Delta Q$ = 0.04–0.09 kWh/kg is required. Table also shows the energy introduced with a jet of a PT operating in our ordinary energetic mode with enthalpy HPL = 3.6 kWh/kg and – for comparisons – in a much less intense mode HPL = 0.72 kWh/kg. One can conclude by comparing the values of $\Delta Q$ and $H_{PL}$ between each other, that in this regime one can confine ourselves to a low-power PT. Otherwise, the excess energy of the plasma torch can be used to vitrify the ash residue. Here it should be taken into account that when working with moist sewage sludge, the energy introduced by the PT is proportional to $\Delta K$ = K - Ko. The introduced thermal energy levels at K = 0.5 exceed the noted values $\Delta Q$ and, in the absence of energy consumption for vitrification, would lead to overheating of the internal volume of the gasifier.

| Parameter | | K, arb. un. | |
|---|---|---|---|
| | | 0.25 | 0.5 |
| Lo, arb. un. | | 0.165 | 0.165 |
| $\Delta Q$, kWh/kg | | 0.04 | 0.09 |
| $Q_{PL}$, kWh/kg | $H_{PL}$ = 3.6 kWh/kg | 0.16 | 0.58 |
| | $H_{PL}$ = 0.72 kWh/kg | 0.03 | 0.12 |
| $PO_2$, kWh | $PO_2$ = 0.35 kWh/Nm³ | 0.04 | 0.035 |
| | $PO_2$ = 1 kWh/Nm³ | 0.114 | 0.1 |
| $k_{NS}$, arb. un. | | 0.934 | 0.807 |
| $W_{SG}$*, kWh/kg | | 5.44 | 5.11 |
| $\eta$, arb. un. | $PO_2$ = 0.35 kWh/Nm³ | 0.015 | 0.023 |
| | $PO_2$ = 1 kWh/Nm³ | 0.028 | 0.033 |

Table: Calculated parameters characterizing the non-stoichiometric process of sewage sludge conversion with its 10% humidity in syngas with oxygen content Lo = 0.165 using plasma technology depending on the amount of water steam K > Ko introduced with plasma jet.

The consumption of electrical energy for the production of a plasma jet with a much lower enthalpy—0.72 kWh/kg is also shown in figure. Without even carrying out detailed calculations, it can be concluded that the use of a less powerful PT would lead to an improvement of the energy efficiency of the process, since it is the level of energy expenditure for the operation of the PT that determines its effectiveness.

The calculated data of table can be useful for assessing the efficiency of the sewage sludge gasification installation, depending on the presence of the mineral mass, which requires vitrification, in its composition. For this, it should be taken into account that at K = 0.5, the next excess energy P is introduced into reactor:

$$P = (Q_{PL} - \Delta Q) / 0.8 = (0.58 - 0.09) / 0.8 = 0.6 \text{kWh} / \text{kg}$$

(when recalculating to electrical energy to power PT). To determine the permissible content of the mineral part in the initial sewage sludge, it is necessary to use the relation $\varepsilon l = 1.48(x/d)^{-0.15}$. If there is a mineral mass in the composition of sewage sludge at a rate of M per 1 kg, the amount of excess energy produced is converted into electric energy, which will be P(1 - M), and it, in turn, can be consumed for vitrification according to $\varepsilon l = 1.48(x/d)^{-0.15}$. Hence we can define M:

$$M(\text{kg}) = 0.35P(1 - M),$$

where the difference in parentheses characterizes the amount of syngas obtained from 1 kg of the mixture. It follows that M = 0.17 kg. Thus, the data of the last column of table for the index of the energy efficiency of the gasification equipment are valid up to 17% of the mineral content in sewage sludge to be vitrified.

A more rigorous problem of the non-stoichiometric gasification regime, self-consistent with respect to energy consumption, is also considered. It was solved on the basis of varying the values of L in the reaction $CH_{2.5}O_{0.5} + KH_2O + LO_2 \rightarrow CO + MH_2 + ECO_2 + DH_2O + Q_{TR}$. for a given value of K. The value of L was determined at which the compensation of the emerging thermal energy deficit ΔQ is attained due to the energy of the plasma jet introduced with the indicated quantity K of water steam at a certain enthalpy. In other words, it was determined at which values of L the condition ΔQ(L) – QPL = 0 is reached.

The main regularities, which ultimately represent the efficiency of the non-stoichiometric gasification process with a small enthalpy of HPL = 0.72 kWh/kg of the plasma jet and in its absence, that is, for wet bottom sludge are shown in figure.

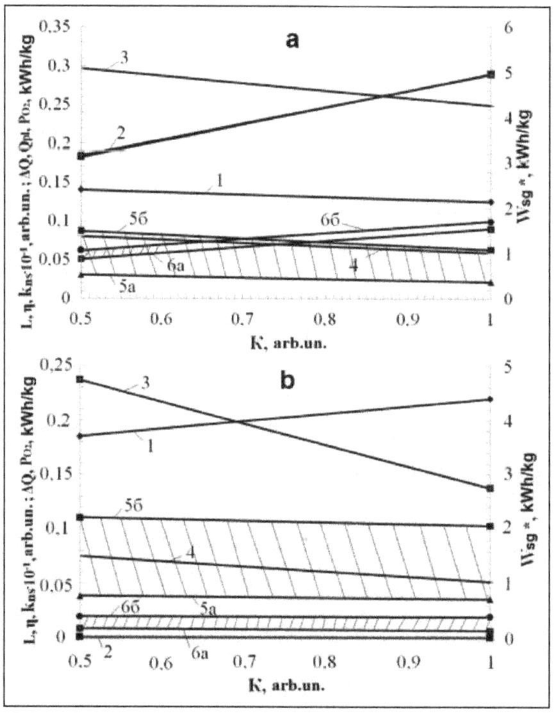


The main regularities characterizing the energy efficiency of non-stoichiometric modes of sewage sludge gasification as a function of the amount of water vapor K introduced into the reaction with the enthalpy of the plasma jet HPL = 0.72 kWh/kg (a), and also, in its absence, for wet sewage sludge (b): 1—the oxygen content L introduced into the reactor; 2—additional energy ΔQ, which should be introduced into the volume to reach the operating temperature, equal to the energy introduced by the steam-plasma jet QPL (the latter—with the exception of wet sewage sludge); 3—the energy of the syngas $W_{SG}^*$; 4—coefficient of nonstoichiometrykNS; 5a and 5b—energy consumption for the production of oxygen at a specific consumption of 0.35 and 1 kWh/Nm3, respectively; 6a and 6b—energy efficiency index of the process at the indicated energy inputs for the production of oxygen, respectively.

In the first of these cases, the energy efficiency index in the entire range of moisture content *K* in the reacting mixture does not exceed 0.1. Even better is the efficiency index of the steam-gasification, that is, in the second case, when its value does not exceed 0.05. However, one should realize that in the reactor space the vitrification and gasification zones are not so separated in space that some of the energy of the plasma jet is not consumed by the gasification processes. Therefore, we believe that, in general, the proposed technology can ensure the energy efficiency of the gasification process for sewage sludge with an index not worse than 0.1.

Thus, practically all cases presented in figure, the consumption of syngas for the electricity generation by means of a gas-diesel power station is only a fraction of the total volume of its production:

$$PO_2 + \Delta Q_{PL} / 0.8) / \eta_{EE} = \eta W^*_{SG} / \eta_{EE} \approx 0.1 \eta W^*_{SG} / 0.3 \approx 0.33 W^*_{SG}.$$

In the variant represented by the last equation, this part is only 30% of the energy for the synthesis gas obtained (in deriving these relations, Equation $\eta = \left(P^C_{PL} + PO_2\right) / W_{SG}$, was used). Accordingly, the remaining part of it can be spent, for example, for the production of electrical energy to external consumers, which will facilitate the commercialization of this development. Thus, in the variant proposed, the processing technology corresponds well to the general idea of numerous publications in the world known as the Waste-to-Energy.

## Incineration

Incineration, or the burning of waste to produce energy is the most direct way to treat and dispose of hazardous waste. It has also been an attractive waste processing approach for many communities around the country. Properly operated incineration projects can provide energy in the form of electricity or processed steam, while reducing the volume of waste that must be landfilled by a significant fraction.

Hazardous Waste Incineration (HWI) is used to describe those facilities specifically designed for treating hazardous waste. Such facilities typically use a rotary kiln design that operates at temperatures above 1100 degrees centigrade along with a highly-advanced flue gas cleaning system. Wherever possible, energy is recovered during the incineration process and used to generate electricity or to produce steam (for either industrial or district heating uses). Hazardous waste incinerators operate to, and comply with, the strictest emission limits of any industry in the European Union, with state-of-the-art facilities combined with experienced and well-trained operators.

HWI plays a prominent role in destroying and eliminating pollutants from hazardous and special wastes. This capacity makes HWI indispensable realising a viable and reliable circular economy.

This is particularly important, as certain hazardous wastes cannot be destroyed by any other means than high temperature incineration. For example:

- Refrigerants containing Chlorofluorocarbons (CFCs), which deplete the ozone layer.
- Polychlorinated biphenyls (PCBs), which do not biodegrade once they enter the food chain, instead building up in fatty tissue.
- Certain cyanide-containing wastes, which accumulate in the environment and pose a risk to human health and ecosystems.

This means that without the option of HWI, some industries could not operate in an environmentally-acceptable manner and as a result would lose their license to operate.

## Physicochemical Treatment

Physicochemical treatment facilities mainly handle liquid hazardous waste, using physicochemical conversion reactions such as neutralization, oxidation or reduction. Prior to such processes, it is common for the hazardous substances to be concentrated through filtration, sedimentation, distillation, ion exchange or the like, so as to improve the process conditions or cost effectiveness of the process in question.

Most physicochemical treatment facilities are based on a technology and operating modality that is specific to the type of waste being treated. For information concerning the state of the art of selected types of physicochemical treatment installations in Germany.

Upon delivery to the waste treatment installation, the target waste undergoes a series of laboratory tests aimed at identifying its contents so that a treatment protocol can be elaborated for the waste. In order to be properly treated, waste usually needs to undergo a series of processes whose nature is determined by the composition of the waste, its chemical reaction behaviour and the desired treatment outcome.

The treatment process is monitored in its entirety, so as to ensure that the maximum amount of pollutants are destroyed or converted.

### Physicochemical Properties

Physicochemical properties are defined as, physical properties, solvation properties related to interactions with different media and reactivity. The physicochemical properties involved in chemical alternatives assessment can be used to identify physical hazards, predict a chemicals environmental fate, human toxicity or Eco toxicity as presented in figure.

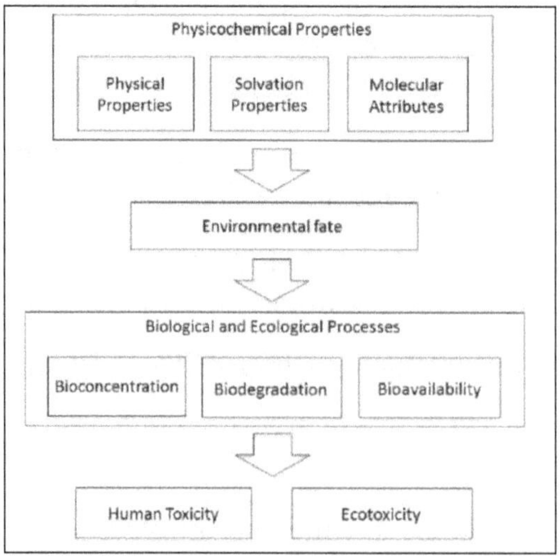


Three primary types of physicochemical properties assessed in this report.

## Physical Properties

Physical properties include freezing point, boiling point, melting point, infrared spectrum, electronic parameters, viscosity, and density. Some of these physical properties (e.g., electronic parameters, molecular weight, boiling/freezing point) are directly associated with environmental fate and health effects.

## Solvation Properties

Solvation properties describe a chemicals interactions with different phases and its partitioning between phases. Solvation properties of interest in alternatives assessment can be divided into three main types:

- Phase partitioning: A partition-coefficient or distribution-coefficient is defined mathematically, as the ratio of concentrations of a given compound across two mixed, immiscible phases at equilibrium. In the context of a chemical alternatives assessment, important partition coefficients are often measured in the liquid phase. Though partitioning can be measured across a range of solvents and phases, the phase partition coefficient most often encountered when, assessing physicochemical properties, is from a system where one solvent is water or an aqueous phase and the second is organic and hydrophobic, such as 1 octanol/water partition coefficient Kow represented by P.
- Solubility: This chemical property refers to the ability of a given substance (the solute) to dissolve in a solvent. The primary measurement of interest in chemical alternatives assessment is solubility in water.
- Colligative properties: Colligative properties are properties of solutions that are not dependent on the chemical species but instead on the ratio of the number of solute particles to the number of solvent molecules in a solution. Examples of colligative properties include lowering of vapor pressure, elevation of boiling point, and depression of freezing point. Colligative properties generally do not play a significant role in alternatives assessments and are not discussed further in this report.

The other properties that was also discuss in this chapter are also molecular attribute, which is used to describe properties related to molecular shape and size and environmental partition coefficients that can provide valuable information about environmental fate such as chemical phase coefficient in soil and water.

## Physicochemical Properties and Environmental Fate

The inherent hazard of a chemical, such as its capacity to intervene in normal biological processes, is the first, while, its physical hazards and environmental fate (degradation time, persistence), is the second. Those outlines, factors are determined by the intrinsic physicochemical properties of the chemicals and the interacting system.

For both organic and inorganic chemicals, those intrinsic parameters are characterized by the molecular structure of the chemical and by the material which is composed. Their hazardousness is then a factor of their size, composition and morphology. The toxicological and physical hazards of the chemicals can be prevented if the physicochemical properties that cause those hazards removed in their initial stage. This procedure is inexpensive but it has to be quickly implemented. Finally, the committee provides additional instructions on the implementation of the framework as presented in figure.

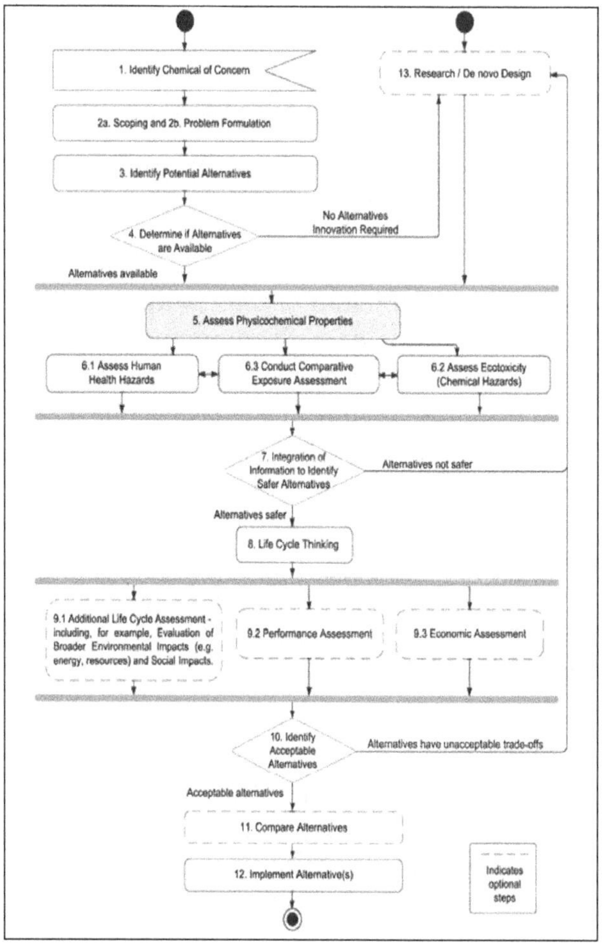


Algorithm representation for assessing physicochemical properties.

## Biological Treatment of Hazardous Waste

Biological treatment uses nature's bacteria and fungi and sometimes costs as little as 10-20% as incineration. Wastes in any form like solid, aqueous and gaseous can be treated biologically. The treatment has six forms. By applying the waste on the soil (land treatment), composting, liquid/solid treatment, using soil to filter gases, in-situ treatment, and conventional municipal/industrial wastewater teatment. The process, and some applications of the first four are described. Views of officials with EPA and the American Petroleum Institute are reported on an upcoming regulation requiring incineration of petroleum refinery waste, unless hazardous components of it are shown not to migrate from the site for as long as they remain hazardous.

Biological treatment methods use microorganisms, mostly bacteria, in the biochemical decomposition of wastewaters to stable end products. More microorganisms, or sludges, are formed, and a portion of the waste is converted to carbon dioxide, water, and other end products. Generally, biological treatment methods can be divided into aerobic and anaerobic methods, based on availability of dissolved oxygen. The purpose of wastewater treatment is generally to remove from the wastewater enough solids to permit the remainder to be discharged to receiving water without interfering with its best or proper use. The solids that are removed are primarily organic but may also include inorganic solids. Treatment must also be provided for the solids and liquids that are removed as sludge. Finally, treatment to control odors, to retard biological activity, or destroy pathogenic organisms may also be needed.

## Biological Treatment Principle

Aerobic self-purification is explained. Three changes occur during aerobic self-purification: coagulation of colloidal solids passing through the primary sedimentation stage; oxidation of carbon, nitrogen, and phosphorus; and nitrification. The basic requirements of any aerobic system for successful treatment of organic matter are a community of acclimatized microorganisms, adequate substrate (food), and a suitable environment. The working principles of waste stabilization and aerobic and facultative ponds are outlined. Three zones can be defined in facultative ponds: a surface zone where aerobic bacteria and algae exist in a symbiotic relationship, an intermediate zone that is partly aerobic and partly anaerobic (usually on a time basis) in which the decomposition of organic matter is carried out by facultative bacteria, and an anaerobic bottom zone in which accumulated solids are decomposed by anaerobic bacteria.

## Biological Treatment System

Biological treatment systems have been used to successfully treat wastewaters from the coking process, even though these wastewaters normally contain significant amounts of toxic substances. Two procedures are used to overcome the toxic effects of these substances. The first is to remove some of them by use of a pretreatment step. The second is to employ the process of gradual acclimation of the biological treatment system to the wastewaters.

The removal step involves combining wastewaters from light oil recovery, final cooling, air pollution control blow-down, and excess flushing liquor. This combined wastewater stream is then passed through the free leg of the ammonia still. The waste from this process is then increased in pH by adding lime. It is then passed through the fixed leg of the ammonia still, where ammonia is removed and recovered. The effluent from this process is then combined with wastewater from the crystallizer and held in a storage tank for a period of time. The pH of the effluent from this holding tank is then lowered with acid to the neutral range. The pretreated wastewater is now ready for treatment by a properly acclimated biological system.

The acclimation step involves subjecting the biological treatment system, activated sludge, for instance, to gradually increasing levels of the toxic substances in the full strength wastewater, until full-strength wastewater is being treated. One way to accomplish acclimation is to develop the activated sludge system using domestic wastewater until it is fully functional as a wastewater treatment system. Then, the pretreated industrial wastewater is added to the domestic wastewater starting with a mix of about 10% pretreated industrial wastewater and 90% domestic wastewater. When it has been established that the treatment system is working well on the 10%/90% mixture, the proportion of pretreated industrial wastewater is increased, and so on, until the treatment system is receiving and operating well on full-strength pretreated wastewater. What happens during the acclimation procedure is that the microorganisms develop the capability to produce enzymes that can metabolize the toxic materials in the pretreated industrial wastewater and that can grow and flourish in the activated sludge (or other biological treatment medium). Those that cannot gradually die off are replaced over time. The result is an "acclimated" biological treatment system that is able to successfully treat the pretreated industrial wastewater.

An acclimated biological treatment system consisting of three extended aeration basins in series has been developed. Phenols are removed in the first basin, oxidation of both ammonia and cyanide takes place in the second basin, and the third basin (nonaerated) is used to remove nitrogen via the denitrification process. It is necessary to maintain favorable concentrations of other nutrients, such as phosphorus, in each of the biological treatment basins.

Chemical-physical treatment systems have been used with some success. One version of chemical-physical treatment has been to use chemical oxidation to destroy organics, ammonia, and cyanide and then to use activated carbon to remove unreacted substances as well as partially treated substances. Oxidation is accomplished by adding chlorine to the pretreated wastes in an aeration basin. Oxygen from air and chlorine are the oxidizing agents. Spent activated carbon can then be recovered (partially) by the process of heating (incinerating).

Biological treatment is extremely effective in removal of nitrate in an anaerobic system. The oxygen receptors prefer nitrate, which is then converted to nitrogen gas and released to atmosphere. If designed correctly, the process will remove almost 100% of the nitrate. However, it may also end up generating some taste and odor problems associated with the anaerobic systems. This is especially true if there are significant amounts of sulfate in the source water which can be reduced to hydrogen sulfide and give the water a rotten egg odor. Additionally, with any biological system, there will be certain metabolic byproducts that could also generate taste and odors and/or complicating chlorination byproducts once the water is disinfected, if chlorine is used. Furthermore, as with any biological system, its success is dependent on keeping the anaerobic organisms in adequate numbers to be effective. Sometimes different chemicals, such as, super-saturated oxygen, contained in the source water may upset the biological system and fail to reduce the nitrates to the level where they should be.

Biological treatment in the subsurface involves the use of microorganisms to break down hazardous organic compounds into non-hazardous materials. The site hydrology, environmental conditions, and the biodegradability of the contaminants are factors that determine the potential effectiveness of in situ biological treatment. Most compounds are more rapidly degraded aerobically, however some compounds will only degrade under anaerobic conditions. Biodegradation in groundwater and solids can be a slow process and may take several years for completion depending on the compounds present. In situ biodegradation, however, is a desirable method of treatment because the contaminants are destroyed, thus, removal of groundwater for external treatment and residual handling possibly can be avoided.

In situ biorestoration of the subsurface is a relatively new technology that has recently gained considerable attention. Scarcely more than a decade ago, conventional wisdom assumed that the subsurface below the root zone of plants was, for all practical purposes, sterile. Research during the last decade has indicated that the deeper subsurface is not sterile, but in fact, harbors significant populations of microorganisms. Bacterial densities of around a million organisms per gram of dry soil have been found in several uncontaminated aquifers. Water-table aquifers examined so far exhibit considerable variation in the rate of biodegradation of specific contaminants and rates can vary two or three orders of magnitude from one aquifer to another or over a vertical separation of only a few feet in the same aquifer. Although extremely variable, the rates of biodegradation are fast enough to protect groundwater quality in many aquifers.

Although not clearly defined, several environmental factors are known to influence the capacity of indigenous microbial populations to degrade contaminants. These factors include dissolved oxygen, pH, temperature, oxidation-reduction potential, availability of mineral nutrients, salinity, soil moisture, the concentration of specific contaminants, and the nutritional quality of dissolved organic carbon in groundwater.

Natural biorestoration does occur in the subsurface environment. Contaminants in solution in groundwater, as well as vapors in the unsaturated zone, can be completely degraded or transformed to new compounds. Undoubtedly, thousands of contamination events are remediated naturally before the contamination reaches a point of detection. On the other hand, methods are needed to determine when natural biorestoration is occurring, the stage the restoration is in, whether enhancement of the process is possible or desirable, and what will happen if natural processes are allowed to run their course.

Biological treatment processes in wetland systems depend on supplying colonies of micro-organisms with optimum quantities of air and nutrients to achieve the same reactions that occur in natural self-purification processes, at a maximum benefit to cost ratio. The three changes that occur during self-purification include coagulation of colloidal solids passing through the primary sedimentation stage; oxidation of carbon, hydrogen, nitrogen, and phosphorus; and nitrification. The basic biological treatment processes used in the system include waste stabilization ponds and constructed wetland systems, trickling (or percolating) filter systems, and activated sludge systems. An aerobic stabilization pond is a large and shallow excavation in the ground, where the treatment of the waste occurs by natural processes involving the use of both bacteria and algae. In aerobic ponds, oxygen is supplied by natural surface re-aeration and by algal photosynthesis. Higher animals such as rotifers and protozoa are also present in the pond. Their main function is to predate on the bacteria, and to a lesser extent on algae, which helps in controlling the suspended solids (SS) concentration in the effluent. Ponds in which the stabilization of wastes is brought about by a combination of aerobic, anaerobic, and facultative bacteria are known as facultative stabilization ponds. The three zones in such ponds include a surface zone where aerobic bacteria and algae exist in a symbiotic relationship, an intermediate zone that is partly aerobic and partly anaerobic in which the decomposition of organic matter is carried out by facultative bacteria, and an anaerobic bottom zone in which accumulated solids are decomposed by anaerobic bacteria.

## Industrial Biosorption Processes

Biological treatment systems where live microbial sludge or biofilms serve the function of metal sorbers have been used in many places for collection and removal of heavy metals from industrial metal-containing effluents. Such systems, however, operate on the principle of bioaccumulation and are not geared to the recovery of metals.

Biosorption systems use the biosorbent materials derived from metal-sequestering biomass. They can operate at elevated metal concentrations which could be toxic to a living system. Industrial biosorbents are either reusable or those which can be disposed off after they are fully loaded and saturated with the sequestered metals. The metal-laden biosorbentts can be shipped to a metal recovery facility where they are either regenerated for reuse or destroyed by ashing with simultaneous metal recovery. In some cases they may be simply disposed off if the sole purpose of the biosorption is remediation. Regeneration of biosorbents with subsequent metal recovery can also be done on the location of the sorption plant.

In commercial operations different types of biomass materials are processed into suitable biosorbents by more or less proprietory procedures. The design and operation of the biosorption process equipment and system do not differ much from that used in the technology of ion exchange adsorption. The process operation is usually based on a contactor system with a bed of either fixed using a normally downflow stream of liquid or fluidized by a normally upflow stream of liquid. Pretreatment of the liquid mainly to remove suspended solid particles may be necessary in some operations. The choice of the biosorption system depend mainly on the amount of the flow to be handled and on its continuity and composition as well as on the conditions for regenerating the biosorbent. The solid-liquid contact sorption system can be scaled up on a modular basis. Good metal uptake performance and selectivity of new biosorbents coupled with their physical characteristics allow great flexibility in application design, thus facilitating ease of operation ease of operation in a variety of metal-removal situations.

Biosorbents are usually made as granular materials for direct sorption of metals from effluent streams. This eliminates the costly and cumbersome chemical pretreatments, which usually generate metal-enriched sludges for either disposal or reprocessing. Specific data on the performance of individual biosorbents are usually not available as most are proprietary items; but this is not a serious handicap as a choice can be made on the basis of published information on systems which are comparable with those required for treatment. Specific operating parameters are then determined by experimental studies.

Regeneration of biosorbents is accomplished in a two-step process consisting of metal elution stage, resulting in a highly concentrated solution of the specific metal and a simple washing stage to complete the regeneration of the biosorbent for recycle.

Volesky has listed the following major interesting features to be considered in newly developed biosorbent materials:

- High versatility: Most operate under a wide range of pH, temperature, and other conditions specific for solution and process.
- Metal selectivity: If required, a wide range of heavy metals can be sequestered without interference from common alkaline earth metals.
- No dependence on concentration: there is high heavy metal uptake for a wide range of metal ion concentration, $< 10$ ppm to $> 100$ ppm.
- High tolerance for organics: Low levels ($< 5$ g/L) of organic contamination do not affect metal uptake.

## Bioremediation

Bioremediation is a waste management technique that involves the use of organisms such as plants, bacteria, and fungi to remove or neutralize pollutants from a contaminated site. According to the United States EPA, bioremediation is a "treatment that uses naturally occurring organisms to break down hazardous substances into less toxic or non toxic substances".

Bioremediation is widely used to treat human sewage and has also been used to remove agricultural chemicals (pesticides and fertilizers) that leach from soil into groundwater. Certain toxic metals, such as selenium and arsenic compounds, can also be removed from water by bioremediation. Mercury is an example of a toxic metal that can be removed from an environment by bioremediation. Mercury is an active ingredient of some pesticides and is also a byproduct of certain industries, such as battery production. Mercury is usually present in very low concentrations in natural environments but it is highly toxic because it accumulates in living tissues. Several species of bacteria can carry out the biotransformation of toxic mercury into nontoxic forms. These bacteria, such as Pseudomonas aeruginosa, can convert $Hg^{2+}$ to Hg, which is less toxic to humans.

Probably one of the most useful and interesting examples of the use of prokaryotes for bioremediation purposes is the cleanup of oil spills. The importance of prokaryotes to petroleum bioremediation has been demonstrated in several oil spills in recent years, such as the Exxon Valdez spill in Alaska, the Prestige oil spill in Spain, the spill into the Mediterranean from a Lebanon power plant and more recently, the BP oil spill in the Gulf of Mexico. To clean up these spills, bioremediation is promoted by adding inorganic nutrients that help bacteria already present in the environment to grow. Hydrocarbon-degrading bacteria feed on the hydrocarbons in the oil droplet, breaking them into inorganic compounds. Some species, such as Alcanivorax borkumensis, produce surfactants that solubilize the oil, while other bacteria degrade the oil into carbon dioxide. In the case of oil spills in the ocean, ongoing, natural bioremediation tends to occur, inasmuch as there are oil-consuming bacteria in the ocean prior to the spill. Under ideal conditions, it has been reported that up to 80 percent of the nonvolatile components in oil can be degraded within 1 year of the spill. Researchers have genetically engineered other bacteria to consume petroleum products; indeed, the first patent application for a bioremediation application in the U.S. was for a genetically modified oil-eating bacterium.

(a)

(b)

(a) Cleaning up oil after the Valdez spill in Alaska, the workers hosed oil from beaches and then used a floating boom to corral the oil, which was finally skimmed from the water surface. Some species of bacteria are able to solubilize and degrade the oil. (b) One of the most catastrophic consequences of oil spills is the damage to fauna.

There are a number of cost/efficiency advantages to bioremediation, which can be employed in areas that are inaccessible without excavation. For example, hydrocarbon spills (specifically, oil spills) or certain chlorinated solvents may contaminate groundwater, which can be easier to treat using bioremediation than more conventional approaches. This is typically much less expensive than excavation followed by disposal elsewhere, incineration, or other off-site treatment strategies. It also reduces or eliminates the need for "pump and treat", a practice common at sites where hydrocarbons have contaminated clean groundwater. Using prokaryotes for bioremediation of hydrocarbons also has the advantage of breaking down contaminants at the molecular level, as opposed to simply chemically dispersing the contaminant.

## Stabilization/Solidification of Hazardous Waste

Stabilization and solidification (S/S) is a soil remediation process by which contaminants are rendered immobile through reactions with additives or processes. During this process, also called immobilization, fixation, or encapsulation, contaminants may be chemically bound or encapsulated into a matrix. The EPA describes stabilization and solidification as a process that accomplishes one or more of the following:

- Improve handling and physical characteristics of waste.
- Decrease surface area of a waste's mass through which transfer/contaminant leakage can occur.
- Limit solubility of hazardous constituents in the waste.

Stabilization is the general term for a process that transforms contaminants into a less mobile or toxic form, while solidification is a more specific process that treats material to increase its solidity and structural integrity. Solidification does not remove nor degrade contaminants, but prevents their transport by eliminating or significantly hindering their mobility.

This process may be performed either ex-situ or in-situ. Ex-situ solidification involves excavation and backfilling, while the in-situ process requires injection of stabilizing compounds into the soil. Typical solidification agents include Type 1 Portland cement, pozzolans, lime, fly ash, and organic binders such as asphalt. A block of solidified soil is often referred to as a "monolith", or a "monofill" if the block encompasses the entire site. Stabilization/solidification is best suited for soils contaminated with metals, radionuclides, other inorganic compounds, and non- or semi- volatile organic compounds. Volatile contaminants are typically not good candidates for solidification, as the contaminants may evaporate into the atmosphere during the mixing process.

Stabilization/Solidification can be classified into three types of processes:

- Aqueous stabilization/solidification.
- Polymer stabilization/solidification.
- Vitrification (high temperature S/S).

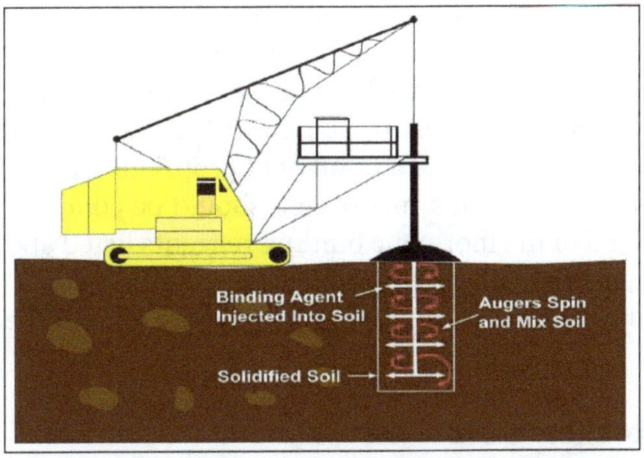


Typical S/S process: the binding agents and wasted soil are mixed with an auger.

USEPA has identified S/S as the Best Demonstrated Available Technology (BDAT) for 57 types of hazardous wastes listed in RCRA.

About 25% of the Superfund remediation sites were treated by S/S technologies.

## Applicability of Stabilization/Solidification Processes

Stabilization/Solidification treatment methods are applicable to a wide range of media including soils, sludges, and sediments (Wilk). Stabilization/Solidification is best suited for soils contaminated with metals, radionuclides, other inorganic compounds, and non- or semi- volatile organic compounds. The S/S process is typically able to encapsulate these contaminants in a matrix and impede their mobility. S/S methods are generally most effective on nonvolatile heavy metals. Volatile contaminants, however, are typically not good candidates for solidification. These contaminants may evaporate into the atmosphere during the mixing process and are difficult to trap within a solidification matrix. The EPA does not currently recognize stabilization as an acceptable treatment method to treat wastes contaminated with polychlorinated biphenyls.

In order to determine effectiveness of solidification for non- or semi- volatile organics, Superfund policy requires a treatability study on similar waste to ensure that the process would decrease the contaminant concentration by 90-99%. For certain constituents, the S/S process performs well at certain concentrations but poorly at others. For example, solidification of copper, lead, and zinc using cementitious materials is effective at low to moderate concentrations. However, these metals interfere with the immobilization process at higher concentrations and render it less effective. The specific mix and distribution of contaminants in the soil also affects the effectiveness of the S/S process.

## Working of Stabilization/Solidification

S/S treatment involves mixing a binding reagent into the contaminated media or waste. Although the terms solidification and stabilization sound similar, they describe different effects that the binding reagents create to immobilize hazardous constituents. Solidification refers to changes in the physical properties of a waste. The desired changes usually include an increase of the compressive strength, a decrease of permeability, and encapsulation of hazardous constituents. Stabilization refers to chemical changes of the hazardous constituents in a waste. The desired changes include converting the constituents into a less soluble, mobile, or toxic form. S/S treatment involves mixing a binding reagent into the contaminated media or waste. Binding reagents commonly used include portland cement, cement kiln dust (CKD), lime, lime kiln dust (LKD), limestone, fly ash, slag, gypsum and phosphate mixtures, and a number of proprietary reagents. Due to the great variation of waste constituents and media, a mix design should be conducted on each subject waste. Most mix designs are a blend of the inorganic binding reagents listed above. Binding reagents that are organic have also been tried. These include asphalt, thermoplastic, and ureaformaldehyde. Organic binding reagents are rarely used in commercial scale due to their high cost compared to inorganic binders.

## Effects of Binding Reagents on Waste

Portland cement is a generic material principally used in concrete for construction. This material is also a versatile S/S binding reagent with the ability to both solidify and stabilize a wide variety of wastes. Portland cement-based mix designs have been the popular S/S treatments and have been applied to a greater variety of wastes than any other S/S binding reagent. Cement is frequently selected for the reagent's ability to (a) chemically bind free liquids, (b) reduce the permeability of the waste form, (c) encapsulate waste particles surrounding them with an impermeable coating, (d) chemically fix hazardous constituents by reducing their solubility, and (e) facilitate the reduction of the toxicity of some contaminants. This is accomplished by physical changes to the waste form and, often, chemical changes to the hazardous constituents themselves. Cement-based S/S has been used to treat wastes that have either or both inorganic and organic hazardous constituents. Mix designs often include byproducts or additives in addition to portland cement. Fly ash is often used to capitalize on the pozzolanic effect of this material when mixed with hydrating portland cement. CKD and slag have minor cementitious properties and are sometimes used for economy. Lime, LKD can be used to adjust pH or to drive off water utilizing the high heat of hydration produced by these S/S binders. Limestone can be used for pH adjustment and bulking.

1. Treatment of Free Liquids: Land disposal of liquid waste or solid-form waste with a free liquid portion is prohibited by RCRA land disposal restrictions. S/S is often used to solidify liquids so that the waste can be land disposed. RCRA policy requires that free liquids be chemically bound. Portland cement is often used as the S/S binding reagent for these wastes since cement reacts with water, chemically binding the water in cement hydration products. An unconfined compressive strength of at least 0.34 MPa (50 psi) is specified to verify that wastes treated for free liquids have had the liquids bound chemically rather than absorbed. This specification is more easily met with the use of cement than other reagents, since the main use of cement in construction is the attainment of compressive strength.

2. Treatment of Inorganic Contaminants: The most popular use of S/S is in the treatment of wastes contaminated with inorganic hazardous constituents. Generally, for inorganic-contaminated wastes, the hazard resides in the heavy metals content. Heavy metal-contaminated wastes are frequently determined to be RCRAcharacteristic wastes due to the leaching potential of the heavy metals. These wastes have failed the toxicity characteristic leaching procedure (TCLP). Frequently, S/S treatment is used to reduce the leaching potential of the hazardous constituent from the waste. After treatment, the waste no longer exhibits the hazardous characteristic (hazardous constituent

leaching) and can be disposed as non-hazardous waste. Many RCRA listed wastes require treatment to the maximum extent practical to reduce their potential hazards when the wastes are land disposed. S/S treatment is often used on RCRA listed wastes to comply with this requirement. At remediation projects S/S is often the only reasonably available technology to treat the large volumes of heavy metals-contaminated soil, sludge, or sediment resulting from these operations. Cement is uniquely suited for use as a S/S reagent for metal contaminants. It reduces the mobility of inorganic compounds by (a) formation of insoluble hydroxides, carbonates, or silicates; (b) substitution of the metal into a mineral structure; and (c) physical encapsulation. S/S treatment can also reduce the toxicity of some heavy metals by changes in valence state.

3. Treatment of Organic Contaminants: The treatment of wastes contaminated by organic hazardous constituents generally relies on cement's ability to solidify the waste. Treatment by solidification relies on changes to the physical properties of the waste. These changes may include the binding of free water in a waste into cement hydration products, creation of waste with more physical integrity such as a granular solid or monolith, and reducing the hydraulic conductivity of the waste. Cementbased S/S treatment has been effective in the treatment for a wide variety of hazardous constituents, including halogenated and nonhalogenated semivolatiles and nonvolatiles, metals, PCBs, pesticides, organic cyanides, and organic corrosives. Treatment of certain organics may require additional attention. Large concentrations of oils and greases (>20%) may prevent the hydration of cement by coating the cement particle with oil or grease thus preventing water from coming in contact with the particle. Some organics can affect the setting time of cement and should be carefully evaluated. Additives and field techniques can often moderate these undesirable effects.

Caution should be taken when using binding reagents that produce significant amount of heat quickly; such as quicklime when mixed with water. This hydration reaction is very exothermic. This fast evolution of the heat can pose challenges in the S/S treatment of materials contaminated with volatile organic compounds (VOCs) and specific other compounds such as polychlorinated biphenyls (PCBs). Air collection and treatment devices may be necessary to avoid transfer of the VOCs from the waste to the atmosphere.

## Physical and Chemical Tests

Most S/S projects require treatability studies and final performance testing of the treated waste. These tests can be placed into two groups: physical and chemical. EPA's publication Stabilization/ Solidification of CERCLA and RCRA Wastes provides descriptions of the various tests used in the United States. It is important to note that the only tests that are required by regulation or policy in the United States are the Toxicity Characteristic Leaching Procedure (TCLP) and the unconfined compressive strength test. Furthermore, these tests are applicable by regulation or policy only in certain limited circumstances. Regulators generally select the appropriate physical and/or chemical tests for a specific project using best professional judgment based on the contaminants and media (soil, sludge, or sediment) and the planned use of the site.

1. Physical Tests: The commonly specified physical tests in project performance standards include the paint filter test (pass/fail), hydraulic conductivity (50 psi)).

2. Chemical Tests: The most commonly specified chemical test is the TCLP, which is frequently applied because it has some relationship to regulations written into the RCRA program. However,

there has been considerable discussion about the appropriateness of applying the TCLP to S/S treated waste when this treated waste is managed other than in a municipal landfill. The TCLP relies on extracting the sample waste with a diluted organic acid (acetic acid), thus simulating conditions of waste co-disposed organic waste such as in a municipal landfill. Many S/S-treated wastes are disposed in monofills or treated and left on site. The TCLP may not be the best simulation of these disposal scenarios. To address this concern, EPA has begun to apply the Synthetic Precipitation Leaching Procedure (SPLP) in lieu of the TCLP. The SPLP (EPA Method 1312-SW846) is designed to simulate waste exposure to acid rain. This procedure is similar to the TCLP except that a weak solution of inorganic acids (sulfuric and nitric acids) is used. Ultimately, project managers and regulators should consider the final disposal environment of the treated waste to determine the appropriate test to use.

## References


- Hazard-waste-treatment-hwt-technologies-physical-and-chemical-treatment-methods-27478: yourarticlelibrary.com, Retrieved 06 April, 2019
- Physical-methods-of-hazardous-wastewater-treatment: nonequilibrium.net, Retrieved 13 July, 2019
- Plasma-gasification-clean-renewable-fuel-through-vaporization-of-waste: waste-management-world.com, Retrieved 13 July, 2019
- Efficiency-of-plasma-gasification-technologies-for-hazardous-waste-treatment, gasification-for-low-grade-feedstock: intechopen.com, Retrieved 30 July, 2019
- Hazardous-waste-incineration: eurits.org, Retrieved 13 March, 2019
- Physicochemical-waste-treatment, waste-disposal, waste-resources: umweltbundesamt.de, Retrieved 30 August, 2019
- Biological-treatment, earth-and-planetary-sciences: sciencedirect.com, Retrieved 06 April, 2019

4

# Disposal and Storage of Hazardous Waste

The process of discarding hazardous waste after its treatment is termed as hazardous waste disposal. Hazardous waste recycling includes glass recycling, plastic recycling, paper recycling, metal recycling, etc. This chapter delves into the methods and related concepts of hazardous waste disposal and storage to provide an in-depth understanding of the subject.

## Disposal in the Normal Trash

Laboratory workers may be surprised to learn the number of wastes they generate that can be disposed of in the normal trash. However, because the disposal of trash from households and businesses is normally controlled by the local municipality, the local agency should be approached to establish what is allowed.

When disposing of chemicals in the normal trash, certain precautions should be observed. Because custodians, who usually empty the trash containers, are not usually familiar with laboratory operations, no objects that could cause harm to them should be disposed of in those containers. Such objects include containers of chemicals, unless they are overpacked to avoid breakage, and powders, unless they are in closed containers. Free-flowing liquids are usually prohibited. Sharp metal and broken glassware, even though they may be considered nonhazardous trash, should be collected in specially marked containers, never in the normal trash baskets.

## Disposal in the Sanitary Sewer

Disposal in the sewer system (down the drain) had been a common method of waste disposal until recent years. However, environmental concerns, the viability of publicly owned treatment works (POTW), and a changing disposal culture have changed that custom markedly. In fact, many industrial and academic laboratory facilities have completely eliminated sewer disposal. Again, like trash disposal, most sewer disposal is controlled locally, and it is therefore advisable to consult with the POTW to determine what is allowed. Yet, it is often reasonable to consider disposal of some chemical waste materials in the sanitary sewer. These include substances that are water-soluble, that do not violate the federal prohibitions on disposal of waste materials that interfere with POTW operations or pose a hazard, and that are allowed by the local sewer facility.

Chemicals that may be permissible for sewer disposal include aqueous solutions that readily biodegrade and low-toxicity solutions of inorganic substances. Water-miscible flammable liquids are frequently prohibited from disposal in the sewer system. Water-immiscible chemicals should never go down the drain.

Disposal of regulated hazardous waste into the sanitary sewer is allowed only in limited situations.

The total wastewater must be a mixture of domestic sewage along with the waste whose amount and concentration meet the regulations and limits of the POTW. If approved of by the local district, it may be allowable to dispose of dilute solutions of metals and other hazardous chemicals into the sanitary sewer.

Under the Clean Water Act, some exemption from regulation as a hazardous waste for wastewater containing laboratory-generated listed waste is allowed. In 1993, this exemption was expanded to include corrosive and ignitable wastes. For the exemption to apply, these laboratory wastes must be 1% or less of the annual total wastewater quantity reaching the facility's headworks or have an annualized average concentration of no more than 1 part per million (ppm) of the wastewater generated by the facility.

Waste should be disposed of in drains that flow to a POTW, never into a storm drain and seldom into a septic system. Waste should be flushed with at least a 100-fold excess of water, and the facility's wastewater effluent should be checked periodically to ensure that concentration limits are not being exceeded.

## Improper Disposal of Household Hazardous Waste

Household hazardous waste (HHW) is not always separated for proper handling before disposal. When disposed improperly to landfills and municipal treatment plants, these products can have significant impact on the environment. Although HHW is a small portion of municipal solid waste, the presence of HHW in solid management facilities that are not equipped to handle them can have problematic effects, resulting in environmental pollution, damage to facilities, and even injury to workers. In many countries, HHW is not subject to legislation unless separated from other household waste because of its small percentage and the challenge in enforcement. In addition, there is no standard definition of HHW globally; therefore, what constitutes to HHW in one country may not be in another. Government legislation and schemes such as Extended Producer Responsibility play a vital role in encouraging proper disposal among consumers, especially when they are convenient and accessible.

Household waste is something that is common among most, if not all, living residences. Like any industrial facilities that handle potentially hazardous materials, households too dispose and use hazardous substances. The chemical complexion in the waste substances makes it so if disposed improperly, it could ignite, explode, poison, or corrode. Household hazardous waste (HHW) becomes what it is once thrown away. Methods of the waste being improperly disposed is pouring the substance down the drain, into storm sewers, on the ground, and throwing it in among the trash. It may not be obvious that these substances, once disposed, will be a danger, but particular varieties of HHWs have the prospective to:

- Cause somatic injury to sanitation workers.
- If poured down drains or toilets, adulterate septic tanks or wastewater treatment systems.
- Pollute—if poured down storm sewers—bodies of water.
- Become a danger to young or unknowing children and pets if left open in the house.
- Contaminate ground and surface water that is used as a way of obtaining drinking water, if directed to exposed landfills.

A big problem that occurs/can occur through improper disposal of HHW would be the deconstruction that the sewage treatment plants are able to obtain. These plants are not able to deconstruct HHW compounds that people would drain or flush, which will end up traveling into lakes and rivers, unprocessed. As a result, one of the main releasers of dioxins and furans was from sewage systems. The substances proved to threaten human health due to the fact that they were highly carcinogenic. Other than the fact of the carcinogenic dangers, interference with the treatments plants could transpire. The toxins that would be processed could poison the microorganisms in the biological process. That would bring us to the position where our water systems would be more susceptible to harmful contaminants.

As a given, hazardous waste is poisonous to all life forms, exposure of such hazardous substances to any living organism (plants and animals) could devitalize it. As a consequence, to the environment, hazardous waste could diminish natural resources and be contaminating to humans. Giving the young/fetuses, whether human or animal, exposure to these hazards would be substantially dangerous, as they are in a process or rapid growth. Introduction to chemicals for the living body would also interfere with biological structure, causing malfunction of organs and limbs.

In addition to the effects to the human and animal bodies, hazardous waste would hinder plant growth. The impeding of plants that are of much use to the human race through manufacturing and consumption would affect our habitat. If the plants were slowly changing, for the worse, it would affect the animals that are needed for food, farm work, and would cause a whole new era of extinction.

If our plant growth can affect our way of living easily, dumping the HHW into landfills gives us a much bigger problem. Landfills that are improperly maintained are major problem; even if they seem to be isolated from any contact, they can contaminate the environment around them. These landfills produce foul-smelling and toxin gases. Along with the gases and toxins, landfills generate leachate, which can travel to our water sources of lakes, rivers, and oceans. This would dig us into a deeper problem of both environmental and human existences. Thereby, leaving HHW unattended and improperly disposed could potentially destroy the ecosystem.

## Household Hazardous Waste

Separate management of HHW from nonhazardous waste is rare. It is estimated that in countries within the Organization for Economic Cooperation and Development (OECD), household waste contributes to 67% of 540 million tons of municipal solid waste (MSW). The estimated amount of HHW varies considerably due to an unclear definition of what constitutes to household waste as opposed to MSW. In the USA, for the Environmental Protection Agency (USEPA), household products that contain corrosive, toxic, ignitable, or reactive ingredients are considered to be HHW. In general, the HHW is a solid, semisolid, or nonaqueous liquid that can cause or significantly contribute to potential hazard to human health or environment when it is improperly treated, stored, transported, disposed of, or otherwise managed. The portion of HHW in MSW has been estimated to be from less than 0.01–3.4% in several studies. The large variability is due to lack of standard definition as to what constitutes HHW, variability in generation, variability in weighing methods, and limited sample size. Nevertheless, 1% by weight is widely accepted as the fraction of HHW in MSW. Because of this small percentage of HHW produced, households are not practically considered to be hazardous waste producers. While HHW represents a relatively small proportion of current urban solid residues, it is the most toxic part of the waste stream.

HHWs in the household waste are often excluded from management as hazardous waste unless collected separately. However, if these waste materials were generated industrially or commercially, they would be subject to strict disposal guidelines. As a result, HHWs are handled the same way as nonhazardous material with no specific regulation or monitoring. Of recent, this mismanagement constitutes a greater problem as the waste stream not only increases in amount but also becomes more diverse with the introduction of more products into the consumer market.

At the source or point of generation, HHW can be placed in the garbage, down the drain, dumped on the ground, or diverted for reuse, energy recovery, or recycle. No matter where HHW is disposed, due to its toxicity as well as municipal treatment facilities that are not equipped to deal with hazardous material, improper management can adversely impact the quality of the environment:

- Contaminate ground water bodies.
- Contaminate surface water bodies.
- Pollute air.
- Affect the human health (children and pets if left around the house, cause physical injury to sanitation workers).

On the other hand, in many third world countries, solid waste management facilities are underdeveloped and sometimes nonexistent. The United Nations reported that between 20 and 80% of all household waste that is generated is often dumped in open spaces, water bodies, drains, and burnt or buried. This creates unsanitary environments leading to health hazards. The portion of HHW in household waste generated by developing countries is much less than in developed countries. The small amount of HHW produced as well as unavailability of funds to direct toward implementing sound practices for waste management has led the United Nations Environment Program to suggest HHW with MSW for disposal in landfills. Regardless of the development level of the country, proper management of HHW can be achieved by understanding the environmental and societal impact of poor practices, HHW contaminants, government legislations, and well-developed schemes.

## Potential Risks of Improper Disposal of HHW

Unavailable facilities for proper HHW management discourage even their voluntary participation. While the products in the HHW list vary from country to country, below are categories that represent majority if not all products that can be classified as HHW:

- Photochemicals.
- Pesticides.
- Mercury-containing wastes.
- CFC-containing equipment.
- Nonedible oil and fat.
- Paints, inks, resins, and adhesives.
- Detergents.

- Pharmaceuticals.
- Batteries.
- Waste electrical and electronic equipment.
- Wood preservatives.
- Aerosols.
- Personal care products.

The risks that a hazardous product poses to the environment depend on certain characteristics of the toxic compounds:

- Solubility.
- Mobility.
- Persistence.
- Degradability.
- Toxicity to nonhuman target species.
- Potential for penetrating landfill liners.
- Potential to be broken down in wastewater treatment system.

HHW is likely to be disposed of improperly because residents do not always understand the level, effect, and potential impact of toxicity in the products that they use. In figure, the disposal trend of households in the UK is presented after a survey with 400 respondents was carried out. One can observe that the predominant method for disposing HHW in households is into the garbage in spite of the toxicity level. A large portion of photochemicals and pharmaceuticals are discarded down the drain with little regard for the compounds that they contain and the consequences for this mode of disposal.

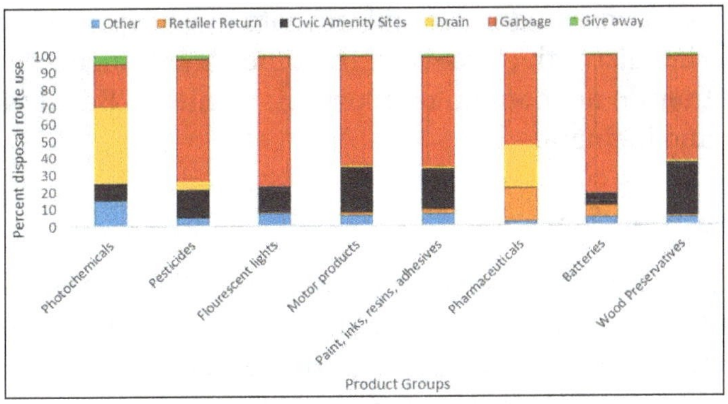


Usual HHW disposal regime of UK households.

Since information about the impact of HHW on the environment is not exhaustive and data relating to disposal are not well known, the potential impact of each of these products in the environment and health is considered as well as the amount that is approximately generated by households where available.

On the other hand, Figure shows a similar study conducted by Statistics Canada in 2009 with over 3800 respondents. While the garbage is still a significant disposal route for HHW, more households reported utilizing drop-off centers and returning products to suppliers and retailers.

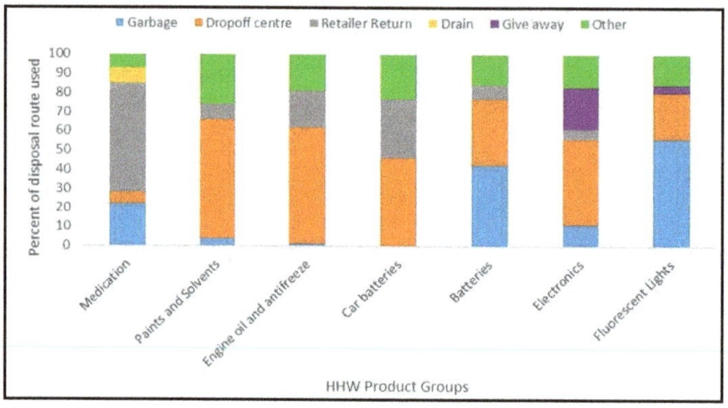


Usual disposal routes in Canadian households.

## Photochemicals

These are liquid chemicals used in home developing and printing. Many of the ingredients in these products are toxic solvents and are predominantly disposed of in sewers. The unused portions of these chemicals are hazardous, but also the packaging can be problematic as it can contain some of the chemicals, which end up in the landfill and thus contaminate both soil and groundwater because these chemicals can penetrate the liners transporting to the groundwater and might end up to the surface water through the movement of groundwater. While the amount entering the sewers cannot be estimated, the packaging in the UK is estimated to be about 270 tons/year, most of which will end up in the landfills.

## Pesticides

Rapid growth in pesticide use has been observed, and this suggests a proportional increase in the amount that is being disposed of. According to the UK Pesticide Safety Directorate, many of the active compounds have been observed in landfill leachate of which research shows that they pose carcinogenic and endocrine disruptive risks. On the other hand, incineration of pesticides is acceptable, provided that they do not contain mercury or arsenic. In Belgium, around 80% of waste pesticides are collected and incinerated.

## Mercury-containing Wastes

Household products that contain mercury include fluorescent bulbs, stockpiled paint, dental amalgam, thermometers, and barometers. Of these, fluorescent bulbs contribute the highest amount of mercury waste. However, as the use of these is reported to have better energy and environmental impacts than regular light bulbs, they are so encouraged. Improper disposal of fluorescent bulbs is where the risks lie. In the UK, it is estimated that 80 million are disposed of each year, of which only a small portion are recycled or processed for mercury extraction. In Brazil, lamps containing mercury contribute 1000 kg of mercury disposed of per year. Mercury exposure poses some health risks such as genetic damage and neurotoxicity damaging the kidney, liver, and central nervous system.

## Chlorofluorocarbon (CFC)-containing Equipment

Refrigeration and air-conditioning appliances/equipment may contain chlorofluorocarbon (CFC) and hydrochlorofluorocarbon (HCFC) refrigerant. CFCs and HCFCs are ozone-depleting substances (ODS). If they released to the environment, they will destroy the protective ozone layer above the earth and potent greenhouse gases, contributing to global climate change. Examples of these types of equipment include motor vehicle and motor vehicle-like air conditioners, central and room air-conditioning units, refrigerators, freezers, chillers, drinking water coolers, dehumidifiers, research equipment, vending machines, etc. Manufacturing of such freezers and refrigerators has been phased out with the CFC component being replaced. However, disposal of these is still ongoing because of their 8- to 12-year life span leaving 4500 tons of CFC in the UK to be safely disposed of. Disposal of equipment that contains ODS is regulated in the EU by the WEEE directive where separate collection is mandatory.

## Nonedible Oil and Fat

Nonedible oil and fat constitutes to about 15% of HHW in the UK. The waste section comprises mineral oils that often contain additives, which make it hazardous. While they are sometimes collected and rerefined or burned for energy, a significant portion is disposed of by end users down the drain or via oil filters and end up in the landfills. There, the oils can disrupt artificial landfill liners. Preferably, the steel component oil filters can be recycled after the oil is pressed for recovery and processed into fuel by companies. Other examples are maintenance lubricants and greases for vehicles, which contain solvents and hydrocarbons that can be just as harmful.

## Paints, Inks, Resins and Adhesives

Disposal of paints is the most significant in this category with the solvent-based paints posing the higher risk. In the UK, paints contribute to 17% of the total HHW with large quantities ending up in the sewers or mixed with MSW. However, schemes developed by local charities exist to collect unwanted paint and redistribute them at no charge. This scheme is limited by the quality of paint that can be used and quantity that can be accepted in any given location. Collected paint needs to meet certain criteria to be acceptable for redistribution such as age or packaging. Barely, 1% of the available excess paint is collected due to a small number of collection points.

## Detergents

The use of detergents in household is widespread. It has been reported that 5–20% of the phosphate that is found in surface and ground water in northern Europe originates from detergent use. However, not all detergents are classified as hazardous, but those containing acids, bases, and chlorinated solvents are of particular concern. In addition, the biodegradability, the aquatic toxicity, endocrine disruptiveness of the surfactants, and other ingredients in the detergent contribute to its classification as hazardous. When combined, some compounds in detergents can release fumes that affect the eyes and mucous membranes, leading to respiratory failure and death after prolonged exposure.

## Pharmaceuticals

In the US, all over-the-counter (OTC) medicines are regarded as hazardous. However, in the EU, only those that are cytotoxic are classified as hazardous. Consumer disposal is not particularly regulated, as it would be problematic, but also due to the relatively low toxicity. As a result of the inability for wastewater treatment plants to remove pharmaceuticals from the waste stream, in many countries, they are now regarded as water contaminants. This is because they eventually make their way into drinking water supplies. They are transferred to sewage sludge during treatment, which is then applied to agricultural land or sent to a landfill.

## Batteries

Primary, lead-acid, and nickel-cadmium batteries are those that fall into this category in HHW making up 6–14% of the HHW in the UK. Mercury in consumer batteries has been banned in Europe and many states in the US. However, many unregulated countries still use batteries containing significant concentrations or mercury, which often ends up in landfills. When buried in landfills, the casing of dry cell batteries can degrade and release heavy metals. Most rechargeable batteries are used in consumer devices and nickel cadmium batteries. In the EU, these types of batteries must be easily removed from electronic devices, and separate collection for recycling is encouraged. However, these end up in MSW where recycling facilities are not well established because it is not mandatory. Cadmium is known to cause health effects like kidney damage. Lead-acid batteries comprise those found in vehicles, or smaller batteries in fire and security alarms. The recycling program for lead-acid batteries in the UK is well established, and 85% collection of the automotive variety has been recorded. However, the batteries from the alarms and from some battery changes carried out at home still end up in MSW. Lead acts as a chronic and acute neurotoxin affecting the kidney.

## Waste Electrical and Electronic Equipment

For many years, home electric and electronic equipment has been disposed of in landfills along with their hazardous components. The amount that is being disposed continues to grow as consumer interest in current devices keeps increasing, which leads to discarding of obsolete electronics. WEEEs often have toxic compounds such that special handling is a requirement. Many countries have prohibited the disposal of WEEE in landfills because of the toxicity and the strain of such large quantities of waste on the landfills. In the EU, this group of equipment is regulated under the WEEE directive such that they are collected and treated as hazardous waste. The directive also lists the substances that should be removed and collected from WEEE. Restrictions have also been placed on the use of certain materials in the manufacture of newer equipment.

## Wood Preservatives

There are three types of treatments that are used to preserve wood, all of which can cause the treated wood to be hazardous, as they have hazardous properties. The types are tar oils, organic solvent-based, and water-based formulations. Creosote, an aquatic contaminant, is often used in tar oils. It is known to be a skin irritant, which causes photosensitivity and skin tumors following long exposure. Tributyltin is an example of organic solvent-based compound that is strictly regulated. A hazardous water-based substance is copper-chrome arsenate (CCA), which contains concentrations of heavy metals that have large health and environmental risks. Arsenate is a priority carcinogenic contaminant of waste, which easily leaches in a landfill and can volatilize during incineration. Landfilling is not acceptable for disposal, and specialized air pollution control equipment is required for incineration.

## Aerosols

Aerosols are a large portion of HHW making up 26% of the HHW in the UK. In the past, CFC was widely used in the production of aerosols. However, CFC has been replaced with alternative propellants and solvents, which contribute significantly to the content in HHW. These replacements are often flammable and explosive. Exposure to aerosols can lead to nausea, skin, and throat irritation.

## Personal Care Products

The harmful nature of PCP has been supported by the discovery of certain long-term effects on health and the environment. While most PCP will end up in the sewers, unused products are stockpiled and end up in MSW.

It is important to understand the fate of compounds in HHW when mixed with MSW for disposal. This has led to stricter disposal regulations in many developed countries to improve HHW management.

## Proper Disposal Methods

## Source Separation

Proper disposal of HHW starts with differentiating between hazardous household products and nonhazardous waste products. Mixing of household waste at the source must be addressed and banned. By collecting similar HHWs together, they can be more efficiently managed with regard to environmental safety, human health, and costs. When separated, arsenic-treated wood can be incinerated using proper pollution control technologies reducing any form of carcinogenic environmental impact, which may otherwise be present if it had been landfilled. Even more popular in developed countries is the separation and collection of cleaning products and pesticides. Majority of which can be incinerated according to best practices unless they contain mercury.

## Recycling and Repurposing

Some HHW products can be of value as they can be recycled for a different purpose or may contain material, which can be extracted for use in manufacturing other products, as in the case of antifreeze, which can be repurposed as an engine coolant. Another example is waste motor oil, which can be refined as lubricating oil or processed as low-grade fuel oil. Lead-acid batteries contain lead, which can be extracted to produce new batteries. Dry cell batteries, on the other hand, contain many different heavy metals, which may pose a problem for extraction. By collecting a significant amount separately, they can be disposed of more cautiously as hazardous waste. Many EU states collect and recycle fluorescent tubes; however, in Germany, all the components of fluorescent tubes aside from the fluorescent powder have been reported to be reused.

## Give Out

Rather than discard surplus products in the garbage or down the drain, items such as paint and wood preservatives can be given out to those who require them when in good condition. Charities that facilitate these have been established in certain countries.

## Methods to Dispose Hazardous Waste

The importance of proper hazardous waste disposal is essential for citizens and business owners alike. Hazardous waste is defined by waste that poses significant or potential threats to the public, health, or environment.

To meet the requirements for hazardous waste, the material needs to display at least one or more of the following hazardous traits: ignitability, reactivity, corrosivity, and/or toxicity. Some of the common Household Hazardous Waste (HHW) is waste that is generated from your own home. These include Paints, Automotive Wastes, Pesticides, Electronics, and Aerosols to Ammunition.

Historically, hazardous wastes were regularly dumped into landfills. This unfavorable action caused the chemicals to seep into the ground and eventually enter our natural water systems. From there, our wild and marine life mammals became exposed to all sorts of chemicals which can in return cause issues for us.

In this day and age of modern technology and electronics, the use of hazardous material is becoming even more prevalent. The only option we are stuck with is to make sure that we dispose of these materials as properly as possible to reduce the risk of harm we create on society.

Here are 5 Techniques and Methods to Properly Dispose of your Hazardous Waste:

- The use of Incineration by burning the material in high temperatures is a great way to destroy toxic waste. Incineration actually destroys and terminates most HHW. A benefit of society using this method is the fact that the flammable wastes can also be burned and used as energy sources. The method of Incineration releases toxic gases which can affect the environment, but current technology has developed more effective incinerator units that limit the amount of emissions released in the sky.
- Another smart method is to recycle. You ever see the cell phone recycling units in Staples

or an Office store. Well, the point of that is to recycle or reclaim the battery and parts off the phone rather than just disposing them. Used oils can also be salvaged and recycled. Recycling is the best method for the environment and is really a "no-brainer".

- Find out what hazardous waste collection facilities are in your area! Many times, communities will designate special days for collection. If you are in immediate need of proper hazardous waste disposal, you can contact your local government or head over to your local municipal building where they should be able to help you.
- If you have diabetic syringes or prescription drugs, you should be very careful about how you dispose of them! Most pharmacies carry a mini disposal box where you can throw away used syringes. Prescription drugs should always be handed off to a drop-off location! They should never be thrown in the trash or flushed down the toilet or sink.
- The last, but not least way to dispose of hazardous waste is to share it. For instance, if you have extra oil or fertilizer lying around you can share it with a person who is need.

By using these methods to dispose of hazardous waste, we can all enjoy a safer environment. Incineration Recycling Services provides all types of industry in the United States with reconditioned industrial packaging containers, as well as empty bulk container pickup services.

## Radioactive Waste Disposal

### Reprocessing

The physical and chemical treatment of spent nuclear fuel to separate the components—uranium, fission products, and plutonium—is given the name reprocessing. The used fuel from the Hanford and Savannah River Plant weapons production reactors and the naval reactors has been reprocessed in the defense program at the US government national laboratories. Commercial experience with reprocessing in the United States has been limited. In the period 1966–72, Nuclear Fuel Services (NFS) operated a facility at West Valley, New York. General Electric finished a reprocessing plant at Morris, Illinois, but design and operational problems prevented it from opening. Another plant was nearly completed by Allied General Nuclear Service (AGNS) at Barnwell, South Carolina, but it never operated as a matter of national policy. To understand that political decision, it is necessary to review the technical aspects of reprocessing.

On receipt of a shipping cask of the type shown in figure, the spent fuel is unloaded and stored for further decay in a water pool. The assemblies are then fed into a mechanical shear that cuts them into pieces approximately 3 cm long to expose the fuel pellets. The pieces fall into baskets that are immersed in nitric acid to dissolve the uranium dioxide and leave Zircaloy hulls. The aqueous solution from this chop-leach operation then proceeds to a solvent extraction process. Visualize an analogous experiment. Add oil to a vessel containing salt water. Shake to mix. When the mixture settles and the liquids separate, some salt has gone with the oil (i.e., it has been extracted from the water). In the plutonium and uranium recovery by extraction (PUREX) process, the solvent is the organic compound tributyl phosphate (TBP) diluted with kerosene. Countercurrent flow of the aqueous and organic materials is maintained in a packed column, as shown in figure. Mechanical vibration assists contact.

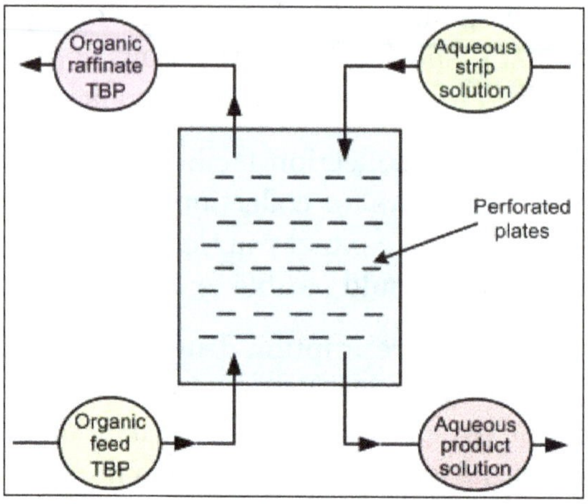


Solvent extraction by the PUREX method.

A flow diagram of the separation of components of spent fuel is shown in figure. The amount of neptunium-239, half-life 2.355 d and parent of Pu-239, depends on how fresh the spent fuel is. After a month of holding, the isotope will be practically gone. The three nitrate solution streams contain uranium, plutonium, and an array of fission product chemical elements. The uranium has a U-235 content slightly higher than natural uranium. It can either be set aside or re-enriched in an isotope separation process. The plutonium is converted into an oxide that is suitable for combining with uranium oxide to form a mixed oxide (MOX) that can comprise part or all of the fuel of a reactor. Precautions are taken in the fuel fabrication plant to protect workers from exposure to plutonium due to its radiotoxicity.

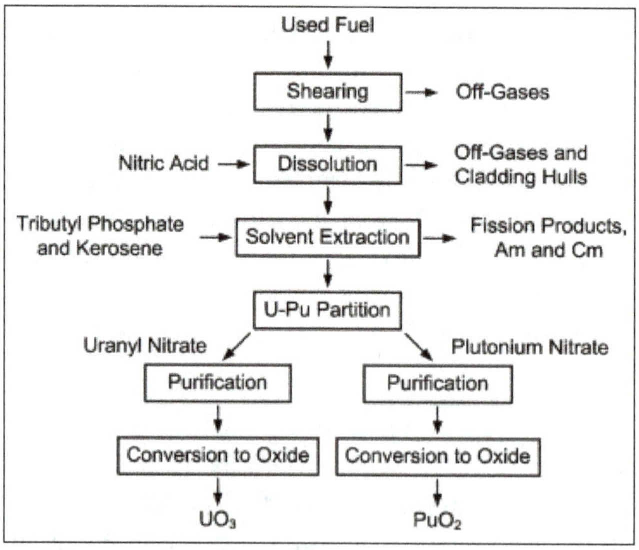


Simplified flowchart of PUREX nuclear fuel reprocessing.

In the reprocessing operations, special attention is given to certain radioactive gases. Among them are 8.04-d I-131, 10.76-y Kr-85, and 12.32-y H-3, which is the product of the occasional ternary fission into three fragments. The iodine concentration is greatly reduced by reasonable holding periods. The long-lived krypton poses a problem because it is a noble gas that resists chemical combination for storage. It may be disposed of in two ways: (1) release to the atmosphere from tall stacks with subsequent dilution or (2) adsorption onto porous media such as charcoal maintained at very low temperatures. The hazard of tritium is relatively small, but water containing it behaves as ordinary water.

Reprocessing has merit in several ways other than making uranium and plutonium available for recycling:

- The isolation of some of the long-lived transuranic materials (other than plutonium) would permit them to be irradiated with neutrons, achieving additional energy release and transmuting them into useful species or innocuous forms for purposes of waste disposal.
- Numerous valuable fission products such as krypton-85, strontium-90, and cesium-137 have industrial applications or may be used as sources for food irradiation.
- The removal of radionuclides with intermediate half-lives allows canisters of wastes to be placed closer together in the ground because the heat load is lower.
- Several rare elements of economic and strategic national value can be reclaimed from fission products. Availability from reprocessing could avoid interruption of supply from abroad. Examples are rhodium, palladium, and ruthenium.
- The volume of wastes to be disposed would be lower because the uranium has been extracted.
- Even if it were not recycled, the recovered uranium could be saved for future use in breeder reactor blankets.

Several countries—France, the United Kingdom, Germany, Japan, and the former USSR—have working reprocessing facilities and benefit from some of the preceding virtues. Analyses by Bunn et al. indicate that uranium prices must significantly increase in order to economically justify reprocessing and recycling of plutonium.

An important aspect of reprocessing is that the plutonium made available for recycling can be visualized as a nuclear weapons material. Concern about international proliferation of nuclear weapons prompted US President Jimmy Carter in April 1977 to issue a ban on commercial reprocessing. It was believed that if the United States refrained from reprocessing, it would set an example to other countries. The action had no effect because the United States had made no real sacrifice, having abundant uranium and coal reserves, and countries lacking resources saw full utilization of uranium in their best interests. It was recognized that plutonium from high-burnup power reactor operation was undesirable for weapons because of the high content of Pu-240, which emits neutrons in spontaneous fission. In addition, it is possible to achieve weapons capability through the completely different route of isotope separation yielding highly enriched uranium. The ban prevented the AGNS plant from operating. US President Ronald Reagan lifted the ban in 1981, but industry was wary of attempting to adopt reprocessing because of uncertainty in government policy and lack of evidence that there was a significant immediate economic benefit.

Alternative proliferation-resistant recycling methods, which should make plutonium less accessible to rogue nations or terrorists, are under development. As a substitute for the PUREX process, which creates a pure plutonium product, the UREX+ process will include other actinides: neptunium, americium, and curium. The mixture of elements will be recycled to burn out wastes and obtain maximum energy.

# Treatment Methods used in Disposal

## Landfill

- Common Landfill – This method involves waste mixing with soil, infiltration and evaporation or shallow burial. Mostly solid wastes are disposed in landfills but slurries, liquids and sludge also might be buried in landfill. The potential damage by landfill caused on environment depends on quantity and composition of waste. Hazardous wastes that are brought in plastic containers or drums contain arsenic, cyanides and such heavy metal compounds and these are accepted in very less quantity by landfill owners.

- Approved landfill – Such sites are appropriate for the organic decomposable materials and inert solid disposal. These sites are situated away from water table to avoid any chance of leaching into water sources.

- Secured Landfill – A securely designed proper landfill site have harmful impact on environment. A site that like this that allows disposal of solid and liquid waste must ensure that in no case it percolates into the surface water or ground water. The main objective is to isolate the landfill site from surrounding environment i.e. water and air quality must be maintained. Some wastes are disposed in open landfills and some in closed landfills, depends upon the waste characteristics.

    - Open Landfills – Wastes like glass bottles and metal cans are incorporated in these.

    - Control Landfill – Wastes that are potentially harmful or toxic such as sludge from sewage or incinerator ash are buries here.

    - Closed Landfills – Wastes that highly toxic that contain mercury compounds or biphenyls where the landfill site is set aloof by a concrete wall, floor and a roof, thus minimizing risk of water percolation into it.

Secure landfill raises the concern of leaching into ground water and to mitigate such risks, USEPA have put forward the solution of using LCRS (Leachate Collection and Removal system) and Double Liners. Liner material is selected keeping in mind the chemical suitability with gas generated from landfill, hydraulic conductivity, stress-strain attributes, durability, aging, etc. thus Darcy establishing a formula relating flow rate, area and driving force to calculate the infiltration for three different liners:

- Soil liner, $q\left(\text{in } m^3/s\right) = \frac{1}{4} K_s iA$
- Geo membrane Line, $q\left(\text{in } m^3/s\right) = \frac{1}{4} K_s iA$
- Composite liner, $q\left(\text{in } m^3/s\right) = 0.21 n^{0.9} a^{0.1} k_s^{0.74}$, for good contact or $= 1.15 n^{0.9} a^{0.1} k_s^{0.74}$, for poor contact.

Where,

Ks = hydraulic conductivity of soil (m/s).

A = Area of soil ($m^2$).

I = hydraulic gradient [(h + d)/d]; h = liquid head (m); d = soil thickness (m).

a = Area of hole ($m^2$).

CB = 0.6 (flow coefficient).

N = porosity = volume of voids/total volume.

G = Acceleration due to gravity ($m/s^2$).

Leachate collection efficiency for landfills is the percentage of sum of liquid collected and that removed in LCRS to total liquid that has been fed to LCRS. Results have shown that efficiency of composite liners are far better mitigating any leaching over the compacted soil.

## Disposal in Adjacent Water Bodies

Mostly solid wastes are stored for temporary period of time in lagoons for minimal impact on environment. Neutralization is done and suspended particles are allowed to settle and for rivers, such wastes are attempted to be released during high flood time. Sometimes, it is directly disposed of on ground.

## Lagoon

It is treated as a temporary as well as permanent storage component even though it leads to potential environmental issues that result from it related to runoff into surface water and leaching into subsurface. Like, during the rains, the lagoon might overflow and thus contaminate the surface.

## Subsurface Leaching

In this method, solid waste is reduced in size and mixed with water to form slurry which is then pumped into cavities that are underground. As the slurry passes through different strata of the soil, it gets treated naturally. There are some new systems that include gravelless or chamber system.

## Waste Initialization before Disposal

Due to the potential risk of the wastes, it should be segregated and recorded. Sometimes, chemical fixation, neutralization, encapsulation and other methods are required. Due to concentrated hazardous wastes, maximum capacity to withhold such wastes is reached. Leachate treatment method is even more complicated due to variety of types of waste and its constituents. Sometimes, due to volatile nature of the hazardous wastes, it needs to be covered immediately and often need extra sophisticated design and care than normally done to municipal waste.

There are ways to pretreat waste before being disposed and these include volume reduction, chemical stabilization, degradation, waste segregation and encapsulation:

- Volume Reduction: Incineration is the best method to achieve this, upto 60% reduction is volume. Major wastes can be totally destroyed except small amounts.
- Chemical Stabilization: It involves addition of chemicals to reduce the movement of

pollutants that help in stabilization and solidification of hazardous waste. It is a cheap method where chemical are mixed with waste sludge thus resulting mixture is pumped into the land where it takes few weeks to solidify. Some of these methods helps in forming matrix and while some get bounded chemically.

- Degradation: Some methods are being utilized for chemical degradation of wastes like oxidation, de chlorination, hydrolysis, etc but no single one line method can be effective in this. Hydrolysis proves to be effective method of all degrading carbamate and organophosphorus pesticides. Using strong oxidants is one approach towards degradation while biological degradation of these in soil is one such method.
- Waste Segregation: Waste is segregated based on chemical characteristics which is usually followed to avoid any unnecessary reaction in the landfill. Number of issues can arise due to mixing of wastes. Segregation before disposing the wastes allows to understand it compatibility with the available facility such as a major acidic waste can used to neutralize a high pH waste.
- Encapsulation: Some wastes which cannot be detoxified may be confined in a material like molten asphalt, concrete and plastics before being disposed. Wastes containing heavy metals should follow such method to avoid leaching of potential toxic materials into the water. For small inorganic waste quantities like one ton of waste per day, steel drums are a good option. In wet regions, full landfill area is covered by addition of asphalt caps or clay to landfill liners but in case of dry regions, encapsulation is not required.

## Selection Criterion for Chemical Waste Landfill

Such landfill sites take benefits of the geological factors in consideration and overcome any disadvantage through its design. The main considerations include.

- Physical Aspects:
    - General topography of the place is taken into account.
    - Presence of water table and geological features of the soil.
    - Examine the presence of streams and surface water.
    - Stability of land and seismic activity in the region.
    - Wind flow and its direction in the area.
- Ecological Aspects:
    - Natural habitat of the area.
    - Conservation/Restrictions by State Government.
- Land Features:
    - Transportation feasibility.
    - Land use is commercial/industrial.
    - Any mining activity.

- Logistics value:
  - Availability of transport.
  - Proximity to industrial region.
  - Firefighting and medical facilities availability.
- Human Aspects:
  - Opportunities for employment.
  - Recreational activity.
  - Health Condition and population density.
  - Archaeological and historical importance.

Some guidelines for Waste Management Practices:

- Segregation of waste products into hazardous and non-hazardous avoids to further complicated products.
- Regular audit to improve continuously on the already established production and operations.
- Preventive maintenance to be followed to avoid any unnecessary wastes.
- All the hazardous material should be stored in enclosed rooms to avoid any contact with other chemicals and protect workers from its effect.
- Label every waste container with details of chemical composition and safety procedures clearly mentioned if exposed.
- Only when municipal body and industry operators work in coordination, the impact of waste would be checked.
- Ensuring no contact of functioning open dumps with atmosphere.
- To any extent feasible, pretreatment of waste before disposal would ensure reduced volume and toxicity.

As far as Indian scenario goes Ministry of Environment, Forests and Climate Change along with the Central Pollution Control Boards and State Pollution Control Boards have acted well in place the Hazardous Wastes (Management and Handling) rules, 1989. Even though regulations were in places, a strict vigilance and lack of proactive approach has led to massive disruption. There are some measures being taken up to implement effective waste management plan:

- Identification of hazardous waste producers, quantifying and records of inventory to be maintained.
- Central and State Pollution control boards have been working and inspecting sites to advice industry operators on reduction of waste production.
- Separate hazardous waste management rules aims at addressing the issues and concerns arising from generation to disposal of hazardous waste. It lists all the hazardous waste

generators along with industry size on its official website and also lists the chemicals that are banned by it.

Regular and surprise visit to industry ensures sampling which are further tested in nearest zonal laboratories set up by CPCB and also reflects efficient working towards waste management. Main features of hazardous waste (handling and management) rules, 1989 are:

- Separate categories of wastes are listed along with various processes that generate such hazardous waste products.
- Every generator must hold an authorization letter to handle hazardous waste which is to be renewed periodically. The authorization can be denied and accepted only after thorough inspections are conducted on sites.
- Records of generation, disposal and storage must be maintained and accordingly file annual return as required.
- Transportation Emergency Card must be provided to the transporter of hazardous wastes.
- State Pollution Control Board records the annual returns and prepares the impact assessment to be submitted to CPCB.

Major focus areas are the following:

- Toxic Substances: The main problem arises from the toxic chemicals being used in the production processes which is leading to secondary formation of furans and dioxins that are major organic pollutants. Working conditions of incinerator is important to avoid the release of secondary contaminant. In India, majority of municipal wastes are burnt while some incinerators that work, doesn't function in the required temperature. Shortage of enough check on incinerators that not only burn MSW but only biomedical, pesticide and petrochemical wastes. Even after sufficient training and awareness of banned pesticides, it is still used and piled up at sites, this increased audits are required.
- Concerns for achieving sustainable development in waste management: Development meant industrialization, urbanization and increased production but it also meant resource depletion and degradation of ecosystem which are made concern in achieving sustainable development. Disintegrated approach of development and environment lead to imbalance in the ecosystem. But now a single framework is being promoted that merges proper technology like biotechnology and vermicomposting in one space. It has been mandatory to utilize full utilization of fly ash produced in boilers and furnaces by brick manufactures. The most recent has been banning plastic bags and all such plastic packaging and encouraging recycled products.

## Hazardous Waste Recycling

A hazardous secondary material is recycled if it is used or reused (e.g., as an ingredient in a process), reclaimed, or used in certain ways including used in a manner constituting disposal and burned for energy recovery.

A material is reclaimed if it is processed to recover a usable product or if it is regenerated (e.g., regeneration of spent solvents).

A material is used or reused if it is either employed as an ingredient in an industrial process to make a product (e.g., distillation bottoms from one process used as feedstock in another process) or if it is employed as an effective substitute for a commercial product (e.g., spent pickle liquor used as a sludge conditioner in wastewater treatment).

"Use constituting disposal" is recycling that involves the direct placement of wastes or products containing wastes (e.g., asphalt with petroleum-refining wastes as an ingredient) on the land. "Burning for energy recovery" is recycling that involves burning a hazardous waste for its fuel value (either directly or when it is used to produce a fuel).

## Benefits of Recycling Hazardous Waste

Hazardous waste reuse, recycling, and reclamation can avoid environmental hazards, protect scarce natural resources, reduce the nation's reliance on raw materials and energy and provide economic benefits.

## Environmental Benefits

There are several, interrelated environmental benefits of recycling hazardous waste including:

- Reducing the consumption of raw materials,
- Reducing pollution,
- Reducing energy use and
- Reducing the volume of waste that must be treated and disposed of.

The extraction, refining, transportation and processing of new raw materials can have a significant impact on the environment. Recycling hazardous waste can mean less air, water, and soil pollution associated with these practices.

Also, recycling can reduce emissions of greenhouse gases (GHGs). When hazardous wastes are recycled, less energy is needed to obtain raw materials and to manufacture products. When energy demand decreases, fewer fossil fuels are burned and less GHGs are emitted into the atmosphere, which can help lessen the impacts of climate change and decrease air pollution.

Finally, by recycling hazardous wastes, less hazardous waste is sent for treatment and disposal. This means less need for hazardous waste landfills and incinerators, as well as a decrease in energy used for those systems, which leads to less pollution.

## Economic Benefits

Hazardous waste recycling is not only good for the environment, but can benefit an organization's bottom line. Recycling hazardous waste can increase production efficiency and reduce costs associated with purchasing raw materials and waste management. By recycling hazardous materials, a business may be able to eliminate the generation of hazardous waste and avoid RCRA regulatory requirements altogether. A business may also benefit from the positive or "green" image associated with hazardous waste recycling efforts For example, a company who values strong corporate stewardship can increase goodwill with shareholders and consumers and further distinguish itself from its competitors.

## Glass Recycling

Glass recycling is the processing of waste glass into usable products. Glass that is crushed and ready to be remelted is called cullet. There are two types of cullet: internal and external. Internal cullet is composed of defective products detected and rejected by a quality control process during the industrial process of glass manufacturing, transition phases of product changes (such as thickness and colour changes) and production offcuts. External cullet is waste glass that has been collected or reprocessed with the purpose of recycling. External cullet (which can be pre- or post-consumer) is classified as waste. The word "cullet", when used in the context of end-of-waste, will always refer to external cullet.

To be recycled, glass waste needs to be purified and cleaned of contamination. Then, depending on the end use and local processing capabilities, it might also have to be separated into different colors. Many recyclers collect different colors of glass separately since glass retains its color after recycling. The most common colours used for consumer containers are clear (flint) glass, green glass, and brown (amber) glass. Glass is ideal for recycling since none of the material is degraded by normal use.

Bottles in different colors.

Many collection points have separate bins for clear (flint), green and brown (amber). Glass re-processors intending to make new glass containers require separation by color, because glass tends to retain its color after recycling. If the recycled glass is not going to be made into more glass, or if the glass re-processor uses newer optical sorting equipment, separation by color at the collection point may not be required. Heat-resistant glass, such as Pyrex or borosilicate glass, must not be part of the glass recycling stream, because even a small piece of such material will alter the viscosity of the fluid in the furnace at remelt.

Mixed color glass cullet.

Public glass waste collection point for different colors of containers.

## Necessity for Glass Recycling

It is a non-biodegradable waste and comprises a lot of the landfills. Glass cannot be decomposed naturally, although it darkens in color but still remains glass in composition. So its recycling is very necessary. It is not harmful for human health, but it gives a dirty look to the land.

## Process of Glass Recycling

### The Collection

The collection of glass from all the household items, glass bottles and broken glass of windows, doors etc. is the first major step. There should be proper bins for the collection of glass so that the work of collecting it from landfills is reduced and glass can reach the recycling sites that are industries easily. There are proper containers for their collection and we can also employ the use of urban green igloos or clean points.

### Separation into Colors

As different colors of glass are not chemically compatible for each other so, different colors of glass like clear, green, amber etc. are separated from each other.

### Formation of Cullet

The glass is then crushed and processed in mills in its powdered form until finally it is called cullet. It is separated from many contaminating materials like rocks and ceramics. Flint cullet is the most desirable form of scrap for recycling.

### Heating and Melting

The cullet is then put into specially designed large furnace where it is heated giving a specific range of temperature and melting it. It does not have a sharp melting point. The heat resistance of glass plays an important role in determining the process to be used.

### Shaping the Glass

The last step in glass recycling is shaping. The melted cullet can be shaped into various forms depending upon your requirement. Usually it is super cooled to turn into different shapes. It is an important quality of glass.

### Use of Conveyors

The next step which is taken after collection, sorting, melting of glass is to be processed in conveyors to reach their final state. They are actually made up of different parts of the process and give us the finished product. Their aim is to move a bulk of material through the process at an efficient rate.

### Recycling into Glass Containers

Glass bottles and jars are infinitely recyclable. The use of recycled glass in manufacturing conserves raw materials and reduces energy consumption. Because the chemical energy required to melt the raw materials has already been expended, the use of cullet can significantly reduce energy consumption compared with manufacturing new glass from silica ($SiO_2$), soda ash ($Na_2CO_3$), and lime ($CaCO_3$). Soda lime glass from virgin raw materials theoretically requires approximately 2.671 GJ/tonne compared to 1.886 GJ/tonne to melt 100% glass cullet. As a general rule, every

10% increase in cullet usage results in an energy savings of 2–3% in the melting process, with a theoretical maximum potential of 30% energy saving. Every metric ton (1,000 kg) of waste glass recycled into new items saves 315 kilograms (694 lb) of carbon dioxide from being released into the atmosphere during the manufacture of new glass.

A variant of the "Tidyman" symbol, intended to encourage people to recycle glass.

## Recycling into other Products

The use of the recycled glass as aggregate in concrete has become popular, with large-scale research on that application being carried out at Columbia University in New York. Recycled glass greatly enhances the aesthetic appeal of the concrete. Recent research has shown that concrete made with recycled glass aggregates have better long-term strength and better thermal insulation, due to the thermal properties of the glass aggregates. Glass which is not recycled, but crushed, reduces the volume of waste sent to landfill. Waste glass may also be kept out of landfill by using it for roadbed aggregate or landfill cover.

Glass aggregate, a mix of colors crushed to a small size, is substituted for pea gravel or crushed rock in many construction and utility projects, saving municipalities, such as the City of Tumwater, Washington Public Works, thousands of dollars (depending on the size of the project). Glass aggregate is not sharp to handle. In many cases, the state Department of Transportation has specifications for use, size and percentage of quantity for use. Common applications are as pipe bedding—placed around sewer, storm water or drinking water pipes, to transfer weight from the surface and protect the pipe. Another common use is as fill to bring the level of a concrete floor even with a foundation.

Other uses for recycled glass include:

- Fiberglass insulation products,
- Ceramic sanitary ware production,
- As a flux in brick manufacture,
- Astroturf,
- Agriculture and landscape applications, such as top dressing, root zone material or golf bunker sand,
- ·cycled glass countertops,
- filtration media,

Mixed waste streams may be collected from materials recovery facilities or mechanical biological treatment systems. Some facilities can sort mixed waste streams into different colours using electro-optical sorting units.

## Plastic Recycling

Plastic recycling is the process of recovering scrap or waste plastic and reprocessing the material into useful products. Since the vast majority of plastic is non-biodegradable, recycling is a part of global efforts to reduce plastic in the waste stream, especially the approximately 8 million metric tonnes of waste plastic that enters the Earth's ocean every year.

Compared with lucrative recycling of metal, and similar to the low value of glass recycling, plastic polymers recycling is often more challenging because of low density and low value. There are also numerous technical hurdles to overcome when recycling plastic. Materials recovery facilities are responsible for sorting and processing plastics, but have struggled to do so economically as of 2019.

When different types of plastics are melted together, they tend to phase-separate, like oil and water, and set in these layers. The phase boundaries cause structural weakness in the resulting material, meaning that polymer blends are useful in only limited applications. The two most widely manufactured plastics, polypropylene and polyethylene, behave this way, which limits their utility for recycling. Each time plastic is recycled, additional virgin materials must be added to help improve the integrity of the material. So, even recycled plastic has new plastic material added in. The same piece of plastic can only be recycled about 2–3 times before its quality decreases to the point where it can no longer be used.

Recently, the use of block copolymers as "molecular stitches" or "macromolecular welding flux" has been proposed to overcome the difficulties associated with phase separation during recycling.

The percentage of plastic that can be fully recycled, rather than downcycled or go to waste, can be increased when manufacturers of packaged goods minimize mixing of packaging materials and eliminate contaminants. The Association of Plastics Recyclers has issued a "Design Guide for Recyclability".

The use of biodegradable plastics or plastics which can be organically recycled or can be composted in industrial composting is increasing for certain short-lived packaging applications.

Plastic recycling bin in Poland.

## Methods

Broadly, there are two major ways to recycle plastic: (1) mechanical recycling ("chop and wash"), where the plastic is washed, ground into powders and melted, and (2) chemical recycling, where the plastic is broken down into basic components.

Before recycling, most plastics are sorted according to their resin type. In the past, plastic reclaimers used the resin identification code (RIC), a method of categorization of polymer types, which was developed by the Society of the Plastics Industry in 1988. Polyethylene terephthalate, commonly referred to as PET, for instance, has a resin code of 1. Most plastic reclaimers do not rely on the RIC now; they use various sorting systems to identify the resin, ranging from manual sorting and picking of plastic materials to mechanized automation processes that involve shredding, sieving, separation by rates of density i.e. air, liquid, or magnetic, and complex spectrophotometric distribution technologies e.g. UV/VIS, NIR, Laser, etc. Some plastic products are also separated by color before they are recycled.

After sorting, for mechanical recycling the plastic recyclables are then shredded. These shredded fragments then undergo processes to eliminate impurities like paper labels. This material is melted and often extruded into the form of pellets which are then used to manufacture other products. The highest quality purification may be referred to as "regeneration".

## Thermal Depolymerization

Scientists have estimated that the potential commodity value of waste plastic may be in excess of $300 per ton when used in process pathways yielding high-value chemical products or to produce electricity in efficient IGCC (Integrated Gasification Combined Cycle) processes.

## Waste Plastic Pyrolysis to Fuel Oil

Plastic pyrolysis can convert petroleum-based waste streams such as plastics into fuels and carbons.

Given below is the list of suitable plastic raw materials for pyrolysis:

- Mixed plastic (HDPE, LDPE, PE, PP, Nylon, Teflon, PS, ABS, FRP etc.).
- Mixed-waste plastic from waste paper mill.
- Multi-layered plastic.

## Heat Compression

Heat compression takes all unsorted, cleaned plastic in all forms, from soft plastic bags to hard industrial waste, and mixes the load in tumblers (large rotating drums resembling giant clothes dryers). The most obvious benefit to this method is that all plastic is recyclable, not just matching forms. However, criticism rises from the energy costs of rotating the drums, and heating the post-melt pipes.

## Distributed Recycling

For some waste plastics, technical devices called recyclebots enable a form of distributed recycling.

Preliminary life-cycle analysis (LCA) indicates that such distributed recycling of HDPE to make filament of 3D printers in rural regions is energetically favorable to either using virgin resin or conventional recycling processes because of reductions in transportation energy.

## Chemical Recycling

For some polymers, it is possible to convert them back into monomers, for example, PET can be treated with an alcohol and a catalyst to form a dialkyl terephthalate. The terephthalate diester can be used with ethylene glycol to form a new polyester polymer, thus making it possible to use the pure polymer again.

An estimated 60 companies are pursuing chemical recycling as of 2019.

In 2019, Eastman Chemical Company announced initiatives of methanolysis and syngas designed to handle a greater variety of used material.

In 2019, Brightmark Energy in the United States began building a facility to convert 100,000 tons of mixed plastic per into diesel, naphtha blend stocks, and wax; the company plans to expand into building another plant which can process an additional 800,000 tons of plastic per year. The company has said that the economics have a significant margin of safety from price declines.

## Other Processes

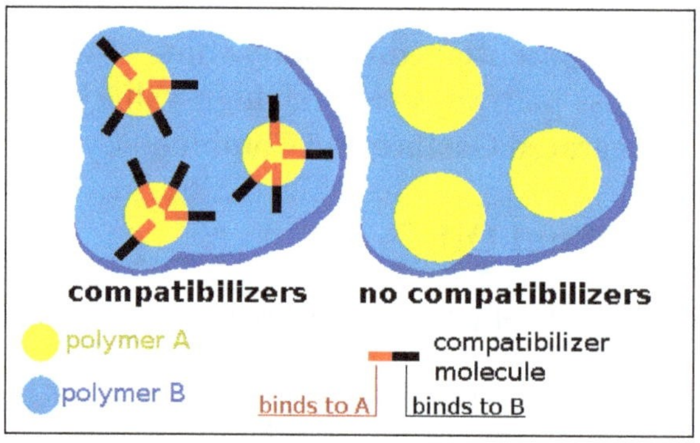


Plastic or other polymer compatibilization.

A process has also been developed in which many kinds of plastic can be used as a carbon source in the recycling of scrap steel. There is also a possibility of mixed recycling of different plastics, which does not require their separation. It is called compatibilization and requires use of special chemical bridging agents compatibilizers. It can help to keep the quality of recycled material and to skip often expensive and inefficient preliminary scanning of waste plastics streams and their separation/purification.

## Applications

### PET

Post-consumer polyethylene terephthalate (PET or PETE) containers are sorted into different color fractions and baled for onward sale. PET recyclers further sort the baled bottles and they are washed and flaked (or flaked and then washed). Non-PET fractions such as caps and labels are removed during this process. The clean flake is dried. Further treatment can take place e.g. melt filtering and pelletizing or various treatments to produce food-contact-approved recycled PET (RPET). This sorted post-consumer PET waste is crushed, chopped into flakes, pressed into bales, and offered for sale.

One use for this recycled PET is to create fabrics to be used in the clothing industry. The fabrics are created by spinning the PET flakes into thread and yarn. This is done just as easily as creating polyester from brand new PET. The recycled PET thread or yarn can be used either alone or together with other fibers to create a very wide variety of fabrics. Traditionally these fabrics are used to create strong, durable, rough products, such as jackets, coats, shoes, bags, hats, and accessories since they are usually too rough for direct skin contact and can cause irritation. However, these types of fabrics have become more popular as a result of the public's growing awareness of environmental issues. Numerous fabric and clothing manufacturers have capitalized on this trend.

Other major outlets for RPET are new containers (food-contact or non-food-contact) produced either by (injection stretch blow) moulding into bottles and jars or by thermoforming APET sheet to produce clamshells, blister packs and collation trays. These applications used 46% of all RPET produced in Europe in 2010. Other applications, such as strapping tape, injection-moulded engineering components and building materials, account for 13% of the 2010 RPET production.

In the United States, the recycling rate for PET packaging was 31% in 2013, according to a report from The National Association for PET Container Resources (NAPCOR) and The Association of Postconsumer Plastic Recyclers (APR). A total of 1.798 billion pounds was collected and 475 million pounds of recycled PET used out of a total of 5.764 billion pounds of PET bottles.

## HDPE

Plastic #2, high-density polyethylene (HDPE) is a commonly recycled plastic. HDPE's highly crystalline structure makes it a strong, high density, moderately stiff plastic. HDPE Thermoplastic materials become liquid at their melting point—around 130 °C. A major benefit of thermoplastics is that they can be heated to melting point, cooled, and reheated again without significant degradation. Instead of burning, thermoplastics like PE (Polyethylene) liquefy, allowing them to be easily extruded or [injection molded] and turned into brand new [HDPE pipe]. Often it is typically downcycled into plastic lumber, tables, roadside curbs, benches, truck cargo liners, trash receptacles, stationery (e.g. rulers) and other durable plastic products and is usually in demand.

## PS

The resin identification code symbol for polystyrene.

Most polystyrene products are not recycled due to the lack of incentive to invest in the compactors and logistical systems required. As a result, manufacturers cannot obtain sufficient scrap.

Expanded polystyrene (EPS) scrap can easily be added to products such as EPS insulation sheets and other EPS materials for construction applications. When it is not used to make more EPS, foam scrap can be turned into clothes hangers, park benches, flower pots, toys, rulers, stapler bodies, seedling containers, picture frames, and architectural molding from recycled PS.

Recycled EPS is also used in many metal casting operations. Rastra is made from EPS that is combined with cement to be used as an insulating amendment in the making of concrete foundations and walls. Since 1993, American manufacturers have produced insulating concrete forms made with approximately 80% recycled EPS.

## Other Plastics

The white plastic polystyrene foam peanuts used as packing material are often accepted by shipping stores for reuse.

Successful trials in Israel have shown that plastic films recovered from mixed municipal waste streams can be recycled into useful household products such as buckets.

Similarly, agricultural plastics such as mulch film, drip tape and silage bags are being diverted from the waste stream and successfully recycled into much larger products for industrial applications such as plastic composite railroad ties. Historically, these agricultural plastics have primarily been either landfilled or burned on-site in the fields of individual farms.

CNN reports that Dr. S. Madhu of the Kerala Highway Research Institute, India, has formulated a road surface that includes recycled plastic: aggregate, bitumen (asphalt) with plastic that has been shredded and melted at a temperature below 220 °C (428 °F) to avoid pollution. This road surface is claimed to be very durable and monsoon rain resistant. The plastic is sorted by hand, which is economical in India. The test road used 60 kg of plastic for an approximately 500-meter-long, 8-meter-wide, two-lane road. The process chops thin-film road-waste into a light fluff of tiny flakes that hot-mix plants can uniformly introduce into viscous bitumen with a customized dosing machine. Tests at both Bangalore and the Indian Road Research Centre indicate that roads built using this 'KK process' will have longer useful lives and better resistance to cold, heat, cracking, and rutting, by a factor of 3.

## Equipment Vendors

Major plastic recycling equipment companies include Tomra. Equipment such as shredders and granulators may be sold by a variety of companies.

In 2016, startup Precious Plastics began selling machines targeted to DIY designers to recycle plastic.

## Recycling Rates

The quantity of post-consumer plastics recycled has increased every year since at least 1990, but rates lag far behind those of other items, such as newspaper (about 80%) and corrugated fiberboard (about 70%). Overall, U.S. post-consumer plastic waste for 2008 was estimated at 33.6 million tons; 2.2 million tons (6.5%) were recycled and 2.6 million tons (8%) were burned for energy; 28.9 million tons, or 86%, were discarded in landfills.

As of 2015, approximately 6.3 billion tons of plastic waste had been generated, around 9% of which had been recycled, 12% was incinerated, and 79% was accumulated in landfills or the natural environment. In 2016 only 14% of plastic waste was recycled globally. According to the EPA, the recycling rate for plastics overall was 9.1% in 2015. Certain products have higher rates, such as PET bottles and jars at 29.9%, and HDPE natural bottles at 30.3%. These rates are lower than certain other materials, like steel cans, that had an estimated recycling rate of 71.3% in 2015.

Japan's plastic waste utilization rate stood at 39% in 1996, increasing to 73% in 2006, 77% in 2011, 83% in 2014 and 86% in 2017, according to the nation's Plastic Waste Management Institute.

## Plastic Identification Code

Seven groups of plastic polymers, each with specific properties, are used worldwide for packaging applications. Each group of plastic polymer can be identified by its plastic identification code (PIC), usually a number or a letter abbreviation. For instance, low-density polyethylene can be identified by the number "4" or the letters "LDPE". The PIC appears inside a three-chasing-arrow recycling symbol. The symbol is used to indicate whether the plastic can be recycled into new products.

The PIC was introduced by the Society of the Plastics Industry, Inc., to provide a uniform system for the identification of various polymer types and to help recycling companies separate various plastics for reprocessing. Manufacturers of plastic products are required to use PIC labels in some countries/regions and can voluntarily mark their products with the PIC where there are no requirements. Consumers can identify the plastic types based on the codes usually found at the base or at the side of the plastic products, including food/chemical packaging and containers. The PIC is usually not present on packaging films, since it is not practical to collect and recycle most of this type of waste.

| Plastic identification code | Type of plastic polymer | Properties | Common packaging applications | Melting- and glass transition temperatures (°C) | Young's modulus (GPa) |
|---|---|---|---|---|---|
| 01 PET | Polyethylene terephthalate (PET, PETE). | Clarity, strength, toughness, barrier to gas and moisture. | Soft drink, water and salad dressing bottles; peanut butter and jam jars; ice cream cone lids; small consumer electronics. | Tm = 250; Tg = 76 | 2–2.7 |
| 02 PE-HD | High-density polyethylene (HDPE). | Stiffness, strength, toughness, resistance to moisture, permeability to gas. | Water pipes, Gas & Fire Pipelines, Electrical & Communications conduit, hula hoop rings, five gallon buckets, milk, juice and water bottles; grocery bags, some shampoo/toiletry bottles. | Tm = 130; Tg = −125 | 0.8 |

| | | | | | |
|---|---|---|---|---|---|
| 03 PVC | Polyvinyl chloride (PVC). | Versatility, ease of blending, strength, toughness. | Blister packaging for non-food items; cling films for non-food use. May be used for food packaging with the addition of the plasticisers needed to make natively rigid PVC flexible. Non-packaging uses are electrical cable insulation; rigid piping; vinyl records. | Tm = 240; Tg = 85 | 2.4–4.1 |
| 04 PE-LD | Low-density polyethylene (LDPE). | Ease of processing, strength, toughness, flexibility, ease of sealing, barrier to moisture. | Frozen food bags; squeezable bottles, e.g. honey, mustard; cling films; flexible container lids. | Tm = 120; Tg = −125 | 0.17–0.28 |
| 05 PP | Polypropylene (PP). | Strength, toughness, resistance to heat, chemicals, grease and oil, versatile, barrier to moisture. | Reusable microwaveable ware; kitchenware; yogurt containers; margarine tubs; microwaveable disposable take-away containers; disposable cups; soft drink bottle caps; plates. | Tm = 173; Tg = −10 | 1.5–2 |
| 06 PS | Polystyrene (PS). | Versatility, clarity, easily formed. | Egg cartons; packing peanuts; disposable cups, plates, trays and cutlery; disposable take-away containers. | Tm = 240 (only isotactic); Tg = 100 (atactic and isotactic). | 3–3.5 |
| 07 O | Other (often polycarbonate or ABS). | Dependent on polymers or combination of polymers. | Beverage bottles, baby milk bottles. Non-packaging uses for polycarbonate, compact discs, "unbreakable" glazing, electronic apparatus housing, lenses (including sunglasses), prescription glasses, automotive headlamps, riot shields, instrument panels. | Polycarbonate: Tg = 145; Tm = 225. | Polycarbonate: 2.6; ABS plastics: 2.3. |

## Asia and Africa

The Ocean Conservancy reported that China, Indonesia, Philippines, Thailand, and Vietnam dump more plastic in the sea than all other countries combined. Scientific American reported that China dumps 30% of all plastics in the ocean, followed by Indonesia, the Philippines, Vietnam, Sri Lanka, Thailand, Egypt, Malaysia, Nigeria and Bangladesh.

## United States

In 2015, the United States produced 34.5 million tons of plastic, which was about 13% of total waste. About 9% of that was recycled. Most of the waste stream is biodegradable but plastic though only 13% of the waste stream is persistent and accumulates.

Low national plastic recycling rates have been due to the complexity of sorting and processing, unfavorable economics, and consumer confusion about which plastics can actually be recycled. Part of the confusion has been due to the use of the resin identification code, which is only found on a subset of plastic products, and which includes the recycling symbol as part of its design. The resin identification code is stamped or printed on the bottom of containers and surrounded by a triangle of arrows. The intent of these symbols was to make it easier to identify the type of plastics used to make a particular container and to indicate that the plastic is potentially recyclable. The question that remains is which types of plastics can be recycled by local recycling centers. In many communities, not all types of plastics are accepted for sidewalk recycling collection programs due to the high processing costs and complexity of the equipment required to recycle certain materials. There is also sometimes a seemingly low demand for the recycled product depending on a recycling center's proximity to entities seeking recycled materials. Another major barrier is that the cost to recycle certain materials and the corresponding market price for those materials sometimes does not present any opportunity for profit. The best example of this is polystyrene (commonly called styrofoam), although some communities, like Brookline, Massachusetts, are moving toward banning the distribution of polystyrene containers by local food and coffee businesses.

## Chemical Recycling of Household Polymeric Wastes

The correct disposal of household polymeric wastes and chemical recycling of household polymeric wastes for chemical feedstock is an important process. The waste generated may cause environmental, economic and health problems. Household waste has lots of polymeric materials. The two most important of the polymers are polystyrene (PS) and low-density polyethylene (LDPE). Main products of PS chemical recycling are: styrene monomer, toluene, ethylbenzene, α-methyl styrene and other valuable chemicals. When LDPE undergoes thermal degradation in a solvent setting in autoclave, oil like diesel can be obtained. Around 6.9% polyethylene terephthalate (PET), 12.1% high-density polyethylene (HDPE), 10.4% polyvinyl chloride (PVC), 17.5% low-density polyethylene (LDPE), 18.9% polypropylene (PP), 7.4% polyurethane (PUR) and 19.7% other polymers are used in the generation of plastic material. Although they differ in polymer, other materials are added mostly in plastic materials as they (polymeric materials) are generated from polymers. The amount of polymer from which plastic material would be generated without adding any material is restricted. Each additive used in the production of plastics has got different purposes except for the ones, such as calcium carbonate, carbon black and caolin, which are used to provide economical production of polymers. These additives are materials such as plasticizers (phthalates), antioxidants, antistatic agents (tertiary amine derivatives, ethylene oxide/propylene oxide copolymers, glycerol derivatives and ethoxylated amines), UV stabilizer (thioethers, hindered amine light stabilizer), antimicrobial agent (10,lO-oxybisphenoxyarsin, copper-bis-(8-hydroxyquinoline), trichloromethyl thiophthalimides), colorants (azo dye, anthraquinone dyes, quinopthalone, benzodifuranones) and blowing agents (chlorofluorocarbons, hydrochlorofluorocarbons, methylene chloride, pentane, cyclopentane and isopentane, azodicarbonamide, azoisobutyronitrile, hydrazine derivatives). Even if these materials are insignificant in plastic generated, it is known that they have toxic, carcinogenic and endocrine deforming chemicals that are harmful to human health. Moreover, it is possible for polymeric wastes in household wastes to cause pollution. The usage of plastic and the disposal of waste following acceptable environmentally sound standards are important.

## Disposal of Household Polymeric Wastes

The disposal of household polymeric wastes differs from one country to the other and also it is not generally possible to collect and eliminate them. In this chapter, chemical recycling of PS and LDPE plastic wastes (polymeric materials) are utilized using three methods: recycling, landfill and energy recovery. 26% of these wastes are regained by recycling, and energy was obtained from 36% by energy recovery and 38% of eliminated by landfill.

## Landfill

In this process, household wastes are neatly stored underground. Landfill process has several disadvantages such as wastes have great volumes, decrease of the areas in which the wastes are stored, high cost and contamination of water and soil. Due to these reasons, the process of landfill should be preferred in a situation when the process of chemical recycling cannot be performed.

## Energy Recovery

In this process, wastes are burned and their energy is used. Even if the disposal of wastes through burning process has advantages in terms of obtaining energy, it has disadvantages because of toxic, carcinogenic and endocrine deforming components generated during this process. Moreover, facilities using burning process are not cost-effective and environmentally friendly.

## Recycling

The main goal is to make plastics (polymers) using mechanical processes (mechanic recycling) or to produce various chemicals or valuable raw materials (chemical recycling) from household polymeric wastes. Even if mechanic recycling has economic advantages, it has been decreased in the molecular weight of polymer and lost of mechanical and physical properties. Therefore, qualified products cannot be manufactured. In other words, there is difference between the material being obtained from the used plastic and the product being obtained from original polymer during mechanical recycling and also it is not processed after definite cycling. In this case, the main solution to eliminate household polymeric wastes is chemical recycling. Although chemical recycling methods may seem economically disadvantaged compared to other methods, in terms of bringing a final solution to waste disposal is important.

## Chemical Recycling of Household Polymeric Wastes

Household polymeric wastes mainly include materials which are made of PET, HDPE, PVC, LDPE, PP, polystyrene (PS) and PUR. The two most important of these polymers are PS and LDPE. PS is used for producing disposable plastic cutlery and dinnerware, CD, smoke detector, license plate frames, plastic model assembly kits, Petri dishes, test tubes, microplates and many other objects. LDPE is used for production of plastic bags, containers, dispensing bottles, wash bottles, tubing, plastic bags for computer components and various molded laboratory.

In recent years, research has been done concerning the management of household plastic wastes by chemical recycling. Thermal decomposition (pyrolysis and thermal degradation) gasification, hydrolysis methanolysis and glycolysis by the type of household polymeric material are the different methods applied in chemical recycling. Thermal decomposition was applied into great part of polymers. During the thermal decomposition, reactive radicals occur in the body of plastic as a result of bond rupture or hain fracturing, and these radicals reveal gas, liquid and solid products as reacting in serial. This method is applied to polymers as pyrolysis and thermal degradation. Pyrolysis is a chemical recycling method. It can be described as degrading a polymeric waste under nonoxygen media (inert, reductive/oxidative etc.) by heat effect. Necessary thermal amount in pyrolysis should be at a level to degrade the chemical structure of polymeric material and to provide the formation of new chemical materials. Polymeric material is fully transformed to solid, liquid and gas products without getting any additives by pyrolysis. The process does not cause air, water and soil pollution. All of the obtained products can be used as raw materials in other industries. There are many kinds of pyrolysis such as batch, semibatch, vacuum, catalytic pyrolysis, etc. Important pyrolysis parameters are heating rate, temperature, medium and time. The medium may be catalytic, inert, oxidative or reductive. According to heating rate, pyrolysis can be classified as slow pyrolysis and flash pyrolysis. The most definite feature distinguishing flash pyrolysis from slow pyrolysis is the stay-duration of products in pyrolysis setting occurring in the splitting and heating rate. Heating rate is in the range of 1–10 °C/min in slow pyrolysis and it is higher than 100 °C/hour in flash pyrolysis. Thermal degradation is the deformation process of polymeric materials as thermal in a solvent setting at respectively lower temperature as pyrolysis.

## Pyrolysis of Polystyrene Waste in Batch Reactor under Inert Medium

Polystyrene can be degraded to the monomer and other valuable chemicals by slow pyrolysis. Schematic representation of the system, including polystyrene plastic wastes' chemical recycling by slow pyrolysis is shown in figure. The system comprises of Pyrex glass reactor, PID controlling oven to heat reactor (1.5 kW), condenser and liquid product gathering container (phase median). Reactor is in diameter of 4.2 cm and at the length of 40 cm. Internal and external temperatures of reactor are measured by Ni-Cr-Ni thermocouple. Plastic waste was located in the bottom of the tube part in a reactor. Reactor was vertically located in high-temperature ceramic oven. Pyrolysis was done in nitrogen atmosphere.

Liquid products of PS pyrolysis are obtained in a condenser located out of reactor and from liquid-receiver plug vessels (phase separator) having salt-ice mixture in its cover added to its output. Moreover, total conversion was calculated by the amount of liquid and gas products. The liquid products are analyzed to illuminate the content of liquid products by Thermo Finnigan DSQ-250 gas chromatography-mass spectrometry (GC-MS). Rtx-5MS capillary column at a length of 30 m and 0.25 mm in internal diameter was used as a column. The analysis terms of GC-MS: injection temperature is 225 °C, temperature of transfer line is 300 °C and helium flow rate is 0.7 ml/min. Temperature program being applied to column was heated to 300 °C at the heating rate of 5 °C $min^{-1}$ after 5-min stand-by at 50 °C and it was waited for 20 min at this temperature (300 °C).

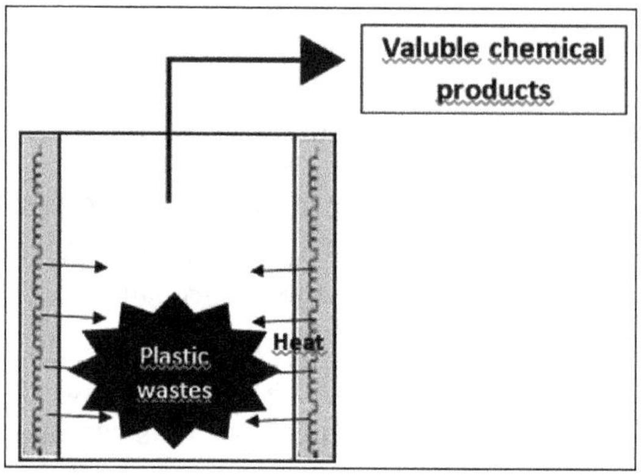


Chemical recycling of plastic wastes via slow pyrolysis.

For chemical recycling of PS polymeric wastes, PS wastes wearere pyrolyzed. The yields wearere investigated in the experiment at 450 °C. Those values are respectively 59.46% liquid, 2.29% solid, 43.46% gas + loss and 92.92% total conversion. According to the results, the majority of the polymer was converted into liquid and gas chemicals. GC-MS chromatogram of the liquid products obtained from chemical recycling of PS polymeric waste is given in figure.

As depicted in table, carbon number ranges of liquid products are found as % 68.8 $C_6$-$C_9$, % 4.36 $C_{13}$-$C_{15}$ and % 26.8 $C_{16}$-$C_{18}$.

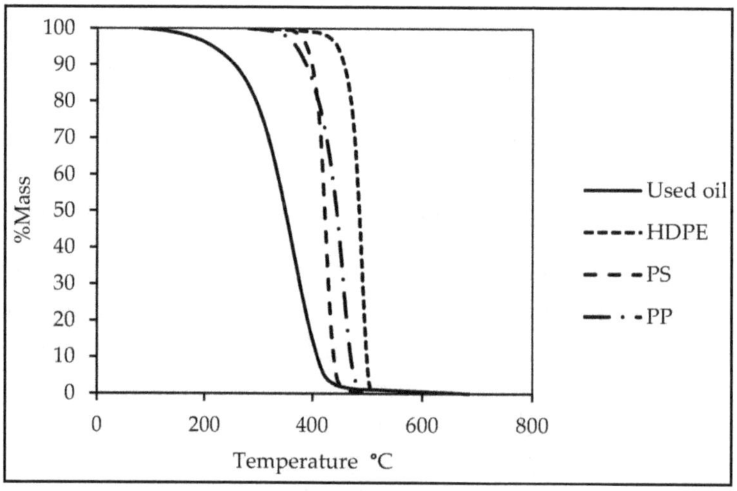


Table: Compounds of PS pyrolysis liquid products at 450 °C.

| Residence time, min | wt% | Compound |
|---|---|---|
| 2.34 | 0.08 | 1-Hexane |
| 2.65 | 0.07 | Bütane-2,3-dimethyl |
| 3.07 | 0.12 | Hexane-3-methyl |

| | | |
|---|---|---|
| 4.23 | 3.62 | Benzene methyl |
| 6.03 | 0.1 | 1-Heptene-5-methyl |
| 6.9 | 0.97 | Ethylbenzene |
| 8.97 | 55.52 | Styrene |
| 11.35 | 4.24 | Alpha-methylstyrene |
| 24.51 | 0.1 | Diphenylmethane |
| 26.56 | 0.37 | Bibenzyl |
| 28.55 | 0.12 | 1,2-Diphenyl cyclopropane |
| 29.7 | 0.12 | Benzene-1,1′(1,3-propanedyl)bis- |
| 31.7 | 22.98 | Naphthalene 1,2,3,4-tetrahydro-2-phenyl |
| 31.91 | 0.09 | Benzene-1,1(1,4-butanedyl)bis |
| 32.34 | 0.1 | Benzene-1,1(1,4-butanedyldiene)bis |
| 32.56 | 0.16 | 1,3-Pentadiene-1,1-diphenyl |
| 32.8 | 0.39 | Benzene, 1,1'-(2-pentene-1,5-diyl)bis- |
| 33.29 | 0.43 | 3,5-Diphenyl-1-pentene |
| 34.21 | 0.09 | Benzene-1,1(1-methyl-2-butylidiene)bis |
| 34.58 | 0.1 | 2,5-Diphenyl-1,5-hexadiene |
| 35.28 | 0.35 | Benzene-1,1(2-pentene-1,5-dyl-bis |
| 36.07 | 0.09 | 1,3-Pentadiene-1,1-diphenyl |
| 36.46 | 0.3 | 1-Pentadiene-1,5-diphenyl |
| 36.85 | 0.11 | 1,5-Diphenyl-1,5-hexadiene |
| 38.87 | 0.08 | 5(2-Propylvinyl)dibenzocycloheptane |
| 42.96 | 0.07 | 1-Ethyl-2-methyl-3-phenylindane |

## Thermal Degradation of LDPE in Autoclave with Solvent

Thermal degradation is a thermal deformation of polymeric wastes in a solvent setting. In this study, LDPE polymeric wastes are put to thermal degradation with vacuum gas oil (VGO) as a solvent supplied from Kirkuk refinery (Iraq). The schema of experiment system where chemical recycling has been done by thermal degradation is given in figure. Experiments are done in a stainless steel reactor called autoclave being durable against high pressure and temperature, gastight, having 250 ml volumes, having pressure and temperature indicator and being screw-capped one. Heating reactor was done by a PID controlling high-temperature oven. Experiments are carried out under nitrogen atmosphere.

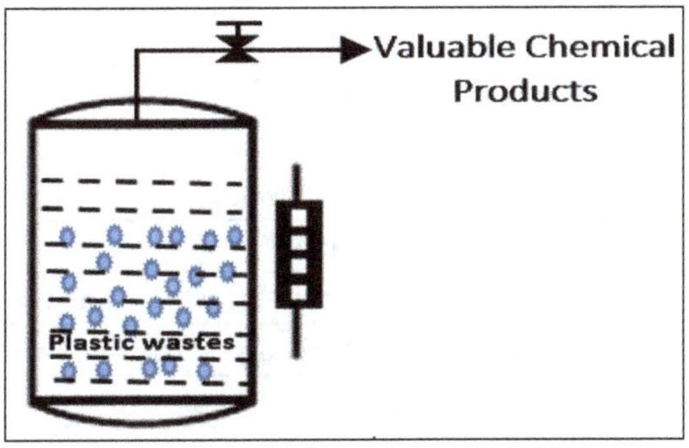


Chemical recycling of plastic wastes via thermal degradation.

At the end of the experiments, autoclave was cooled to room temperature and gas product obtained was taken. Then, the content amount in the reactor was determined as being separated into liquid and solid. The amount of gas product was found from variation in the beginning of process. On thermal degradation, experiments being done as getting LDPE/VGO:1/1, 85.94% liquid, 5.76% solid, 8.3% gas and 94.24% total transformation (liquid + gas) are found.

The content of liquid products for thermal degradation has been illuminated by GC-MS device which is given above. However, the column temperature program has been set at 50 °C at the beginning and after 2-min duration, it has been set at 270 °C at the heating rate of 5 °C/min and it was waited for 20 min at this temperature. GC-MS chromatogram of the liquid products is given in figure. As it can be seen in figure, each pick of the liquid products has tripled. One of them is alkane (saturated hydrocarbon), another is alkene (unsaturated hydrocarbon) and the other is diene (hydrocarbon that contains two carbon double bonds).

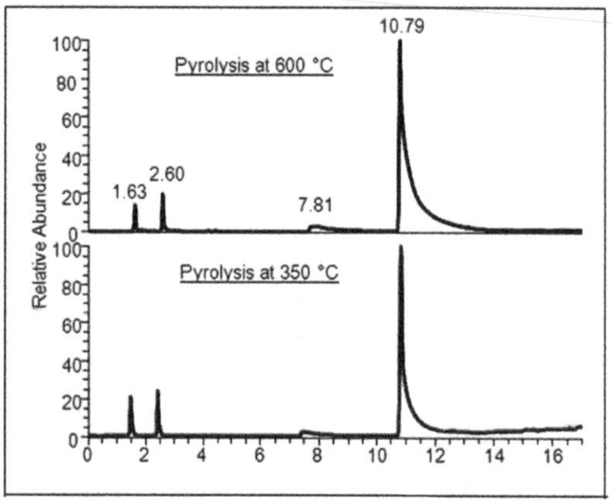


Thermal degradation liquid products obtained from Autoclave are analyzed by using GC-MS. The list of chemicals is given in table.

Table: Compounds of LDPE/VGO:1/1 thermal degradation Liquid products at 400 °C.

| Residence time, min | wt% | Compound |
|---|---|---|
| 1.76 | 5.67 | Pentane |
| 1.99 | 7.46 | Hexane |
| 2.58 | 8.47 | Heptane |
| 2.83 | 1.87 | Methylcyclohexane |
| 3.44 | 3.51 | Toluene |
| 3.84 | 6.55 | Octane |
| 4.43 | 1.71 | Ethylcyclohexane |
| 4.7 | 0.48 | Unknown |
| 5.14 | 3.46 | Ethylbenzene |
| 5.88 | 6.56 | Nonane |
| 6.31 | 0.35 | 2-Methyl-bicyclo-octane C9 |
| 6.58 | 1.24 | 7,7-Dimethyl-tetracycloheptane |
| 7.34 | 0.98 | n-Propylbenzene |
| 7.57 | 1.11 | 1,3,5-Trimethylbenzene |

| | | |
|---|---|---|
| 7.96 | 1.31 | Cyclodecane |
| 8.48 | 5.03 | Decane |
| 8.79 | 0.50 | Cyclodecane C10 |
| 9.27 | 0.84 | Cyclodecane C10 |
| 10.09 | 1.44 | Undecane C11 |
| 11.25 | 5.60 | Undecane |
| 11.65 | 0.36 | Unknown |
| 12.24 | 0.51 | 1-Dodecene C12 |
| 12.57 | 0.54 | 1-Cyclopropyl-1-methyl-benzene C12 |
| 12.87 | 1.06 | 1-Cyclopropyl-1-methyl-benzene C12 |
| 14.04 | 5.36 | Dodecane |
| 15.28 | 0.46 | Dodecane |
| 16.7 | 4.55 | Tridecane |
| 17.04 | 0.38 | Spiro(tricycloundeca-2,4,6-triene),7,1-cyclopropane |
| 17.77 | 0.43 | 7-Tetradecene |
| 19.24 | 3.54 | Tetradecane |
| 20.34 | 0.64 | Unknown |
| 21.63 | 3.10 | Pentadecane |
| 22.79 | 0.71 | Unknown |
| 23.92 | 2.80 | Hexadecane |
| 25.15 | 0.49 | Hexadecane |
| 26.07 | 2.16 | Heptadecane |
| 27.01 | 0.34 | 2,6,10-Trimethyl tetradecane |
| 27.24 | 0.33 | Unknown |
| 28.12 | 1.88 | Octadecane |
| 29.24 | 0.41 | Unknown |
| 30.08 | 1.68 | Nonadecane |
| 31.93 | 1.25 | Eicosane |
| 32.95 | 0.41 | Docosane C22 |
| 33.71 | 0.95 | Docosane |
| 35.39 | 0.73 | Tricosane |
| 37.03 | 0.47 | Pentacosane |
| 38.58 | 0.33 | Pentacosane |

Liquid products of LDPE/VGO:1/1 at 400 °C are examined to evaluate the cetane number of fuel specifications. Cetane number of liquid products from thermal degradation has been found to be nearly 50. This value can be an initiative indicator for this product as it would be used for diesel oil. Consequently, when LDPE is put to thermal degradation in a solvent setting in autoclave, an oil like diesel can be obtained.

## Flash Pyrolysis of Household Polymeric Wastes in Free Fall Reactor

Flash pyrolysis, one of the chemical recycling methods of household polymeric wastes, can be used with various reactors such as fluidized bed and rotating cone. One of these reactors is FFR.

Valuable chemicals can be obtained by flash pyrolysis of household polymeric wastes in FFR. The schema of flash pyrolysis system with FFR is given in figure. Polymeric wastes are made as a granule at a definite size in this system and it is fed suddenly to high-temperature heated reactor by a solid feeding system.

Reactor of FFR experimental system using chemical recycling of PS and LDPE household polymeric wastes is 140 cm in length and in 5 cm diameter quartz. It is heated to pyrolysis temperature with a high-temperature oven of 120 cm length and 10 cm internal diameter. Reactor was fed at 2 g/min feeding rate with a polymeric waste feeding system. There is tar receiver under reactor, serial condenser connected on outlet of tar receivers (which has got salt-ice mixture in its cell) and vacuum pump getting all system under vacuum. Temperature measurements are done in the inner and outer parts of the reactor.

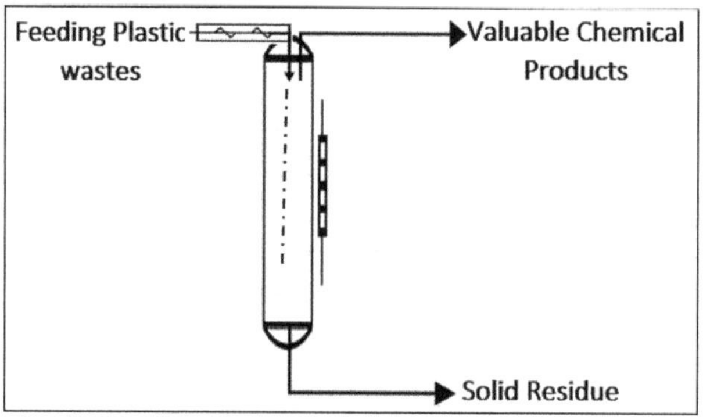


Flash pyrolysis of polymeric wastes in FFR.

Particles of PS and LDPE polymeric wastes at size range of 150–75 micro particle are used in PS and LDPE's flash pyrolysis in FFR. 825 °C flash pyrolysis liquid products are analyzed in GC-MS in conditions of the above given PS and LDPE liquid product analysis both for PS and LDPE in FFR.

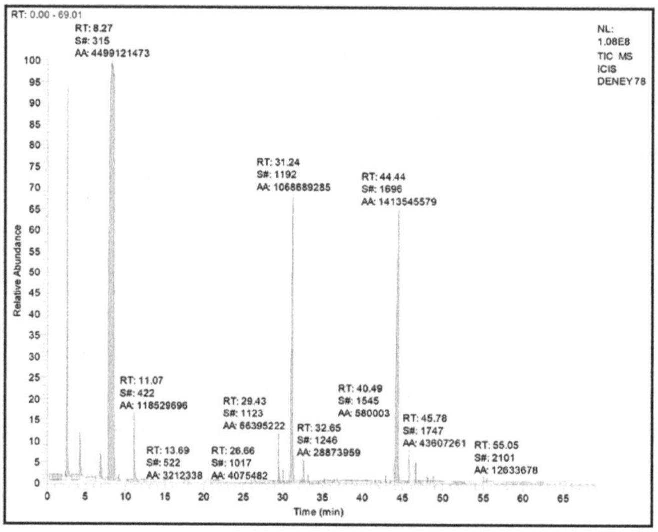


GC-MS chromatogram of PS flash pyrolysis liquid.

GC-MS chromatogram of liquid products from 825 °C flash pyrolysis of household polystyrene polymeric wastes is given in figure. Moreover, the description study of chemical compounds in liquid products was done and the obtained results are given in table.

Table: Compounds of PS flash pyrolysis liquid products in FFR at 825 °C.

| Residence time, min | wt% | Compound |
|---|---|---|
| 4.26 | 1.29 | Toluene |
| 6.85 | 1.05 | Ethylbenzene |
| 8.27 | 53.60 | Styrene |
| 11.07 | 1.41 | α-methylstyrene |
| 29.43 | 0.79 | 1,3-Diphenyl propane |
| 31.24 | 12.73 | 1,1-Diphenyl-2-methyl propane |
| 32.65 | 0.34 | 1,3-Diphenyl-butane |
| 33.23 | 0.21 | 1,1-Diphenyl-2-methyl propane |
| 42.87 | 0.14 | Dimer |
| 44.44 | 16.84 | Trimer (2,4,6-triphenyl-1-hexane) |
| 45.78 | 0.52 | Trimer |
| | 11.60 | Others |

Table depicts that the main product is styrene monomer. Flash pyrolysis of polystyrene in FFR showed that it can obtain important liquid chemicals such as toluene, ethylbenzene, α-methyl styrene and others besides styrene monomer.

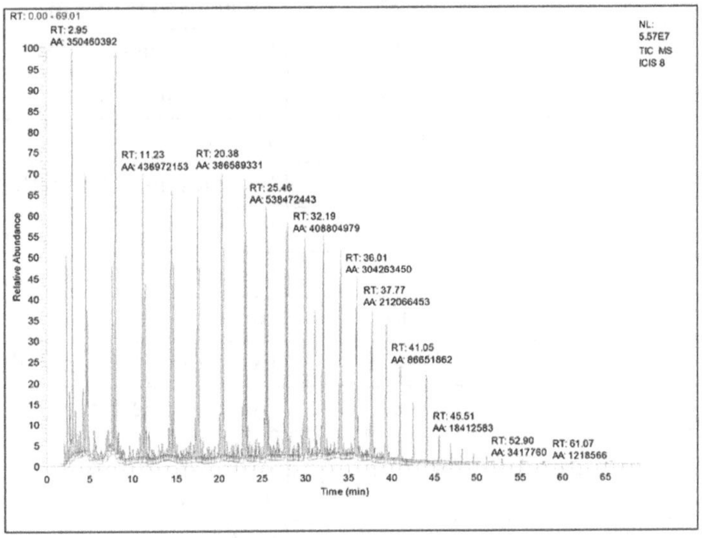


GC-MS chromatogram of liquid products for LDPE wastes in FFR at 825 °C.

GC-MS chromatogram of liquid products for LDPE polymeric wastes in FFR from 825 °C is given in figure. As it can be seen from figure, peaks in chromatogram are given respectively as homologues. This indicates that products have got aliphatic character. LDPE pyrolysis liquid products are described by using GC-MS library.

The products obtained with the use of GC-MS are given in table. As it can be seen in table, the liquid product which obtained LDPE is paraffinic hydrocarbons. Liquid products obtained from chemical recycling of LDPE polymeric wastes at 825 °C in FFR weraree divided into three groups. The percentage of groups, The percentages of the groups are % 30.44 $C_6$-$C_9$, % 42.83 $C_{10}$-$C_{13}$ and % 26.73 $C_{14}$-$C_{18}$.

Consequently, gathering, dissection and innovator chemical recycling processes are important for the method of HHW plastic wastes. Gathering and dissection processes for polymeric wastes in HHW have direct effect on chemical recycling. Thus, polymeric wastes should be collected separately from other wastes without being contaminated and should be distinguished from each other. Each of separated polymeric waste should be put to chemical recycling by a proper process. An innovator chemical recycling process should create a harmful waste for environment, health and economy. Recycling of polymeric wastes is advantageous with pyrolysis. Pyrolysis of polymeric waste is carried out in a closed area, where gas, liquid and solid products are obtained. An innovator pyrolysis process should include units making these products usable as raw materials. The following situation is suitable: to gather polymeric wastes in HHW, to dissect them by polymer and their chemical recycling method by polymer. Pyrolysis is suggested for the recycling of PS polymeric wastes and thermal degradation is suggested for LDPE.

Table: Compounds of LDPE flash pyrolysis liquid products in FFR at 825 °C.

| Residence time, min | wt% | Compound |
|---|---|---|
| 2.37 | 2.06 | 1-Hexane |
| 2.74 | 0.28 | 1,3-Pentadiene-2-methyl |
| 3.03 | 4.15 | 1-Heptene |
| 3.42 | 0.77 | Cyclohexane-1-methyl |
| 3.99 | 0.24 | Cyclopentane-1-methyl |
| 4.28 | 1.06 | Cyclohexane-1-ethyl |
| 4.91 | 6.83 | 1-Octane |
| 5.6 | 0.79 | Cyclohexane-1,2-dimethyl |
| 6.03 | 0.38 | 1-Hexadiene-2,5-dimethyl |
| 6.95 | 0.68 | Cyclohexane-1-ethyl |
| 8.18 | 10.12 | Nonane |
| 8.98 | 0.4 | Cyclohexane(1-methylethyldiene) |
| 9.65 | 0.27 | Cyclopentene-1-butyl |
| 11.33 | 5.49 | 1-Decane |
| 11.57 | 3.48 | 2-Decene |
| 11.69 | 0.33 | Cyclohexene-1-butyl |
| 11.96 | 0.32 | 2-Decyne |
| 12.46 | 0.34 | Bicyclo(3,1,1)heptane-2,6,6-trimethyl |
| 14.63 | 4.88 | Undecane |
| 14.87 | 3.86 | Undecyne |
| 16.73 | 0.51 | 1,3-Di(1-propyl) cyclopentane |
| 17.67 | 4.76 | Dodecane |
| 17.91 | 4.23 | 3-Dodecyne |
| 18.17 | 0.38 | 3-Dodecyne |
| 18.83 | 0.37 | 1,1,2-Tridecadiene |

| | | |
|---|---|---|
| 19.51 | 0.36 | 4-Tridecene |
| 20.43 | 4.83 | 4-Tridecene |
| 20.71 | 4.13 | Tridecane |
| 20.95 | 0.39 | 1,1,2-Tridecadiene |
| 21.73 | 0.37 | 4-Nonene-5-butyl |
| 22.2 | 0.4 | 1,1,2-Tridecadiene |
| 23.04 | 5 | 1-Tetradecene |
| 23.31 | 4.09 | Tetradecane |
| 24.54 | 0.37 | 5-Tetradecene |
| 25.74 | 7.61 | Pentadecene |
| 27.95 | 4.76 | Pentadecene |
| 30.02 | 2.01 | Hexadecane |
| 32.01 | 0.77 | 1-Octadecene |

## PET Bottle Recycling for Sustainable Textiles

Polyethylene terephthalate (PET) is a versatile material and has a wide range of applications such as clothing, acoustic panels, sportswear, agricultural nets, nonwovens, sheets and films, straps, engineering resins, food and beverage bottles, bottles, packaging materials, reinforcement in building construction, etc. Among these products, bottle grade PET is generally used for water and beverage packaging due to its lightweight, inexpensive price, resistance to microorganisms, and light. Bottles of water, soft drinks, and other beverages constitute 83–84% of global PET resin requirement. Furthermore, the projected demand for PET packaging materials is forecasted to reach 20 million tons by 2019 with an annual increase of 4.6%.

With the widespread application of PET, large quantities of PET waste were inevitably created. PET has no side effects on the human body and does not pose a direct threat to the environment. On the other hand, it is regarded as a harmful material because of its high volumetric fraction in the waste stream and high resistance to atmospheric and biological agents.

Due to poor biodegradation of PET, it is difficult to remove waste. It is possible to suggest two acceptable solutions; burning and recycling. Burning method arises releasing toxic fumes into the atmosphere, causing environmental pollution and health risks. As an acceptable solution, the recycling of PET bottles enables the conservation of natural sources such as fossil fuels and energy, solving landfill problem, reducing greenhouse gas emission, lowering carbon footprint, creating new business opportunities as well as a contribution to the national economy. In addition, recycling processes are the best way to economically reduce PET waste. With both reduced energy costs and raw material costs, recycling fiber production has become a form of production with a significant economic advantage. Two forms of PET bottle recycling can be distinguished as a closed loop and open loop recycling. Closed loop recycling or bottle-to-bottle refers to a product system that recycles post-consumer waste within the same system. Open loop recycling denotes the utilization of recycled material in another product system such as bottle-to-fiber recycling. Figure displays the process of bottle-to-fiber recycling.

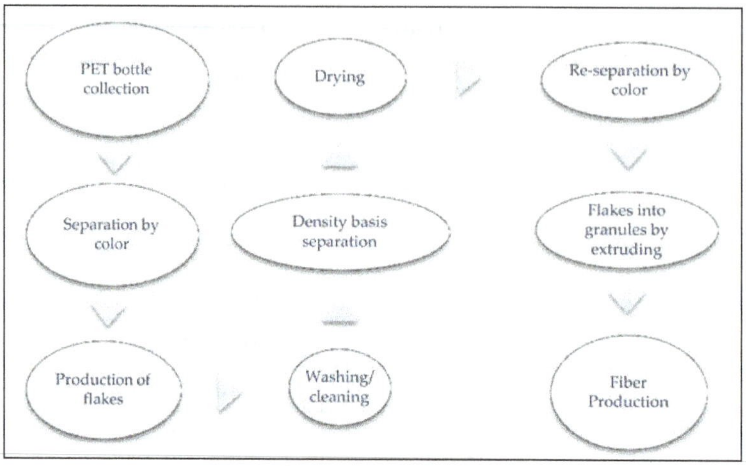


Bottle-to-fiber recycling.

PET flakes are obtained from PET bottle wastes after a series of procedures such as sorting, washing, grinding, drying, etc. Most of the recycled PET flakes produced worldwide are utilized for staple fiber applications in textile sector. Because of environmental reasons initially, the recycling of PET bottles to textile fibers has now become commercially attractive. Furthermore, as petroleum prices increase, recycling of PET becomes more financially feasible rather than a virgin PET.

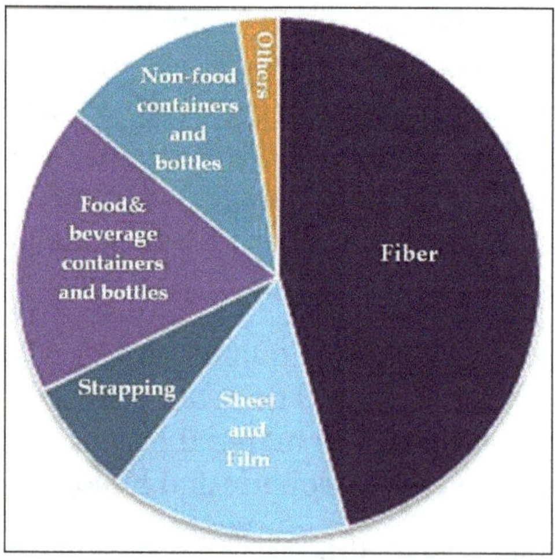


Global annual r-PET market volume by end-use in 2016.

The fiber obtained from PET flakes has dominant proportion among end users as 44% of total market share in 2016. These fiber materials are generally called as recycled PET (r-PET) and especially used in carpets, blankets, clothing, and other textile applications. However, properly sorting PET bottle wastes and carefully removing impurities are essential so as to obtain similar recycled fiber quality as virgin ones. Furthermore, there are significant limitations in the use of r-PET for the production of partially oriented yarn (POY), drawn textured yarn (DTY) and fully drawn yarn (FDY) type textile yarns, microfilaments, tire cord or high quality biaxial films.

Converting of PET flakes into PET fibers can be carried out by two main processes, such as chemical recycling and mechanical recycling. The chemical recycling method provides value-added products from PET bottle wastes, and depolymerization of PET by hydrolysis, methanolysis, and glycolysis is used to re-use regenerated raw materials as monomers for new polymerization processes.

The chemical recycling technique produces superior quality materials, but this method is highly labor and power-intensive, so it requires high processing costs.The mechanical recycling method involves sorting and separation of waste, washing for removal of dirt and contaminants, grinding in order to obtain flakes, cleaning, separating, dehydrating, drying, and re-melting. Mechanical recycling is preferred due to a significant reduction in processing costs, global warming potential, non-renewable energy use, abiotic exhaustion, acidification, eutrophication, human toxicity, and water toxicity.

As such, studies on the use of r-PET staple fiber in the field of textile applications have established the focus of researchers. Researchers investigated the properties of yarns and knitted fabrics produced by virgin PET and r-PET and cotton fibers as virgin and recycled. Yarns were produced as 100% virgin, 100% recycled, and 50%/50% virgin/recycled proportion by using rotor spinning machine. They stated that yarns produced from recycled fibers had better yarn unevenness, lower number of yarn imperfections, and better yarn quality index value. In addition, yarn tensile strength and knitted fabric burst strength were found to be lower for recycled yarns and fabrics when compared to virgin ones.

Another study was conducted on the performance and durability of woven fabrics made from r-PET with different ratios. Commercially available plain and twill woven samples were collected and categorized according to the recycling content. They observed that more fatigue action occurred for fabrics including r-PET than virgin PET after washing. It was found that fabrics with r-PET exhibited higher stiffness with increasing r-PET content. Rajamanickam and Vasudevan studied on the antibacterial activities of 19.7 tex ring spun yarns including lyocell and r-PET at different blending ratios (100% r-PET, 70%/30%, 50%/50% lyocell/r-PET, and 100% lyocell). Before the antibacterial activities, yarns were treated with chitosan finish. In this study, it was suggested that the yarns can be used as hospital textiles and blended yarn samples exhibited better antimicrobial activity. Furthermore, "blends of lyocell & r-PET yarns" were found to be suitable in the field of hospital textiles due to having higher tenacity and elongation property.

Researchers have studied the performance of r-PET and v-PET blended yarns with cotton at different blend ratios (100%, 70%/30%, 50%/50%, and 30%/70%). In that respect, 30 tex yarn samples were manufactured by means of ring spinning system and tensile properties, unevenness, imperfection index (IPI), and hairiness of these yarns were determined. In this study, it is suggested to encourage the consumption of the lower cost r-PET fibers having current quality standards in this field. Another study of the authors is related to the knitted fabric performance made from these yarns. It was found that knitted fabrics including virgin PET had better performance specifications than that of r-PET. According to the test results, they offered to use blends of r-PET fibers in suitable proportions according to usage area in order to prevent the decrease in fabric performance instead of using virgin PET fiber.

In order to contribute the use of r-PET staple fiber in the textile sector, producers should be encouraged and some legislation should be implemented. Since recycling of PET bottles is an important value-added sector; it prevents the consumption of limited fuel source and energy so it helps to protect the environmental pollution and decreases the landfilling problem. There are various attempts of big textile brands about using of r-PET fibers in their products such as Zara, H&M, Nike, Adidas, and Lewis etc. to fulfill their social responsibilities to nature. With this study, we aimed to determine the performance of r-PET containing yarns produced with different spinning systems at the same production conditions. For this purpose, r-PET fibers obtained from PET flakes by mechanical recycling methods were used and blended with cotton fiber at different ratios. The yarn samples with the same blend ratios were produced with v-PET fiber instead of r-PET, in order to compare the performance of recycled and virgin fibers. Two spinning systems were selected as ring and compact which are commercially used in industry. Tensile properties, unevenness, IPI, hairiness of these yarns were determined and compared. IBM SPSS 20 statistical package program was used to determine the statistical significance of the effects of raw material, spinning system and blend ratio on yarn performance characteristics.

## Thermoplastic Recycling

The choice of recycling of waste thermoplastics depends on processing equipment such as injection, single-screw extruder and film blowing machine, and processing conditions (temperature, time, content of materials, and rheological behavior) and product uses. The application of additives or modifiers like compatibilizer (nonreactive and reactive), fillers or fibers (inorganic and organic) have been attributed to ease processing and improvement in compatibility. More so, the recycling of waste thermoplastics is cheaper than virgin types, but its inferior properties, contaminations, and poor suitability remain an issue of concern for effective applications. Blending technology remains a proffer solution due to low cost to produce, lower technical risk, and eco-friendly materials when compared to developing new polymers. Sorting or separation before recycling through manual application of principle of density and solubilization with the use of solvents (hexane, benzene, xylene, and toluene) provides solution to contamination but not cost effect and risk. Techniques for modifications of thermoplastics may be due to the use of different waste or virgin thermoplastics and natural materials, thereby producing composites with enhanced properties and durability. This can be influenced by processing, crystallization, and phase morphology. The use of different waste or virgin thermoplastics seems to be uneconomical due to cost of blended and non-compatibility of the thermoplastics which may require a new compatibilizer. The use of natural materials for modification of virgin and waste thermoplastics remains a potential technique for thermoplastic recyclates. Therefore, the major reasons for modification of plastic resins in the industries include to meet specific processing and performance specification of a plastic product that is not satisfied by a single component, to upgrade the properties of postconsumer plastic wastes, for scientific research, for interest and development, and for financial optimization. However, degradation of thermoplastic materials by chemical processes is a function of reaction between the components and the environment. The reduction in photodegradation of thermoplastics by ultraviolet absorber as an antioxidant shows a retardation effect of oxidation. Therefore, the aging process of thermoplastics can be influenced by the synergistic action of factors like electromagnetic radiation and thermal energy on the oxidation, favoring the initiation of degradation by excision of chain and radicals of thermoplastics.

## Modifications and Properties of Recycling of Virgin and Waste HDPE

The incorporation of the carbon nanotube, zeolite, LDPE, PP, natural fillers, and fibers with treatment into the waste polymer for reuse has resulted to an improvement of the composite strength and enhancement of compatibility of blended components of composites. The improvement has been reported to be a function of compatibilizer types, size and particle shape, branching, and dimensions of polymeric chains as reported by. In the case of natural fibers or fillers, it seems fibers or fillers containing compatibilizer which may or may not have been identified. Moreover, the melting flow rate of recycling of postconsumer or waste HDPE remains inconsistent with stabilization, and the consistency can be achieved with a mixture of phosphite and phenolic. This might be uneconomical. The enhancement in mechanical properties and performance of the HDPE matrix and composite product by additives (sodium hydroxide (NaOH), sodium lauryl sulfate (SLS), and acetic anhydride) have also been attributed to increased interfacial adhesion coupled with its improved water absorption, biodegradability, biocompatibility, antimicrobial activity, and non-toxicity with the use of chitosan compounds. The density of recycled virgin and waste HDPE is within the range of 0.02–0.96 g/cm$^3$. Increase in density can be ascribed to chemocrystallization, annealing effects and changes in lamellar orientation, fiber loading, moisture absorption, and aging of HDPE products. Annealing effect involves changes in spherulite size of HDPE material after heat effect, and aged surface shows loss of gloss observed as a result of environmental effect through oxidative stress and disappearance of crystalline molecule of the HDPE materials produced by a surface contraction. The surface contractions initiate micro-cracks and lead to embrittlement of ductile HDPE polymers.

Table: Mechanical properties of modified virgin and waste HDPE materials.

| Materials | Modification | Tensile strength (MPa) | Tensile modulus (MPa) | Flexural strength (MPa) | Flexural modulus (MPa) | Hardness | Impact strength (J/m$^2$) |
|---|---|---|---|---|---|---|---|
| Virgin HDPE | — | 21 | 189 | — | — | — | 6.8 |
| Virgin HDPE | Using LDPE | 22.5 | 860 | — | — | — | 135 |
| Virgin HDPE | 3% Carbon nanotube and 2 cycles | 36 | 1700 | | | | 5 |
| Waste HDPE | Natural zeolite, clinoptilolite ($K_2$,$Na_2$,Ca) $Al_6Si_3O_{72}$. $23H_2O$ of 1–2% with particle size <40 µm | 21.8 | 218 | — | — | — | 25 |
| Waste HDPE | — | 24.619 | 836.25 | 27.114 | 1390.7 | 21 | 859.3 |
| | Combretum dolichopetalum fiber | 32.427 | 939.6 | 18.2 | 1568.1 | 28 | 496.0462 |
| | Acetic anhydride-treated Combretum dolichopetalum fiber | 38.5153 | 1220 | 8.5111 | 19944.24 | 33 | 787.3806 |
| | NaOH-treated Combretum dolichopetalum fiber | 34.9041 | 984.99 | 32.067 | 2277.15 | 39 | 469.5912 |
| Waste HDPE | — | 27.628 | 792.59 | 34.519 | 1390.7 | 24 | 962.8 |
| | Cissus populnea fiber | 30.4827 | 839.022 | 36.1904 | 1425.89 | 30 | 155.795 |
| | Cissus populnea fiber treated with NaOH | 29.6903 | 793.05 | 39.3962 | 1568.44 | 35 | 398.62 |
| | Cissus populnea fiber treated with SLS | 31.8013 | 823.245 | 39.568 | 1455.68 | 38 | 394.683 |

## Modifications and Properties of Recycling of Virgin and Waste LDPE

The high quantity of waste LDPE and its average mechanical properties coupled with influence of aging of the product have not motivated utilization in many packaging applications such as bags, film, and pallet covers, but modifications may improve the mechanical properties. Also, the qualities of LDPE composites have been linked with poor interfacial adhesion between both phases of individual constituents which explain weak mechanical properties. This interfacial adhesion has a direct relation to compatibility. The processing conditions of machine also influenced the compatibility of the polymers. The use of virgin and waste or recycled PP to modify LDPE using twin and single-screw extruder has been reported by Sylvie and Jean-jacques. PP increases some mechanical properties such as tensile strength and modulus with reduced impact strength of the LDPE for single extruding machine, although the twin extruding machine gave better mechanical properties due to improvement in homogeneity of the polymer. The use of compatibilizer such as EPDM, graft copolymer (PE-g-poly (2-methyl-1,3-butadiene), and ethylene-propylene copolymer enhanced the interaction between the polymers and resilience, thereby improving the mechanical properties of the LDPE/PP composites. The use of compatibilizer in virgin and recycled polyolefins influenced the quality of the composites based on technology and recycled waste of LDPE by the addition of EPDM compatibilizer. The presence of ethylene-propylene diene monomer (EPDM) revealed the variation in properties such as wide-angle x-ray diffraction (WAXD), differential scanning calorimetry (DSC), and mechanical properties of virgin and recycled LDPE/PP. The destruction of thermal and mechanical properties of virgin LDPE and PP as well as blended LDPE/PP was found to be greater than those from recycled polyolefins because of the absence of antiaging in the virgin products. The impact EPDM modifier have been reported on stability of LDPE/PP products based on natural and influenced ageing conditions with improved mechanical (tensile and impact strength) properties of the LDPE/PP with increase in modifier content. The impact EPDM modifier significantly improved the compatibility of recycled LDPE and PP and reduces the recrystallization of PP in the blends during aging and decreases the formation of the imperfect β polymorph crystal which depends on the presence of additives resulting in chain mobility retardation, presence of shear stress changing the chain structures, and fast cooling conditions at foil production. Moreover, the significant improvement in rheological property such as viscosity, crystallinity index, and tensile properties of the recycled LDPE can be achieved by linear low-density polyethylene (LLDPE) blend with a ratio of 4:1 and applicably good for film products at 60% blended LLDPE. The modification of recycled LDPE by PP using injection molding machine was also reported that the tensile properties increases with reduction in impact strength as increase in PP content as well as reduced processing temperature.

Table: Mechanical properties of unmodified and modified virgin and waste LDPE.

| Thermoplastics | Modification | Tensile strength (MPa) | Tensile modulus (MPa) | Hardness | Impact strength (J/m²) |
|---|---|---|---|---|---|
| Virgin LDPE | Starch grafted with maleic anhydride | 16.34 | 520.16 | — | — |
| Virgin LDPE | Virgin PP | 25.1 | | | 5.5–6.5 |
| Waste LDPE | Natural zeolite, clinoptilolite $(K_2,Na_2,\text{-}Ca)Al_6Si_3O_{72}\cdot 23H_2O$ of 1–2% with particle size <40 μm | 19.5–22.9 | 195.73–232.14 | | 18–26 |
| Waste LDPE | Waste 10% PP | 8.7–12.1 | 241–336.1 | | 23–37.2 |

| | | | | | |
|---|---|---|---|---|---|
| Waste LDPE | Waste 10% PP + EPDM | 7.6–8.5 | 211.1–236.1 | | 46.4–53 |
| Waste LDPE | — | 30.33 | 240.7 | 2.3 | 583 |
| Waste LDPE | Husk filler | 31.58 | 565.7 | 13.15 | 600 |
| | Okpa filler | 35.14 | 861.2 | 17.53 | 583 |
| Virgin LDPE | 10% PP using single-screw extruder | 9.4 | 205 | | 15.2 |
| Waste LDPE | 10% PP | 10.0 | 248 | | 12.3 |
| Virgin LDPE | 10% PP using twin screw extruder | 9.6 | 226 | | 8.5 |
| Waste LDPE | 10% PP using twin screw extruder | 10.3 | 256 | | 12.6 |
| Waste LDPE | 10% PP + 5% graft copolymer using twin screw extruder | 11.8 | 280 | | 12.5 |
| Waste LDPE | 10% PP + 5% EPDM using twin screw extruder | 10.1 | 245 | | 16.5 |

The effect of wood flour of *Pinus radiata* as fillers at a constant loading of 45 wt.% of recycled post-consumed plastic waste reported to be influenced by virgin PP. In the report, the addition of virgin PP improved tensile and flexural moduli and flexural strength of wood plastic/LDPE composites (WPC). The highest mechanical properties of recycled LDPE composites have been reported for wood polymer composites with virgin PP (5%) and lower mechanical properties with higher virgin PP content of 55 and 71.5% compared to PE. The moisture absorption of WPC with virgin PP blend reported to be higher than without PP with adverse effect on the mechanical properties when immersed in water. More so, the use of virgin PP delay degradation and lower the thermal stability of WPC. The decrease in tensile strength with increasing starch content in starch/LDPE composites attributed to incompatibility of the hydrophobic LDPE and hydrophilic starch and the increase in stiffness attributed to better dispersion of starch in LDPE matrix. This incompatibility demands the use of compatibilizers such as styrene/ethylene-co-butylene/styrene grafted with maleic anhydride (SEBS-g-MA) and anhydride grafted polypropylene (PP-g-MA), Mixture Irganox 1098/Irganox 1078-Irgafos 168/Chimassorb 944.

The novel application of natural materials (filler or fibers) is to enhance undesirable properties and poor biodegradation of LDPE matrix. The use of rice husk, bambara, and mahogany fillers with improved tensile strength and modulus, flexural strength and modulus, and hardness with reduction in impact strength has been reported. The increase in mechanical, thermal, and biodegradation behaviors of the composites was attributed to improved interfacial adhesion and compatibility. The reduction in impact strength is a result of fiber dispersion, uneven distribution, and micropore formation in the composites. It can be deduced that natural fillers or fibers contain a compatibilizer which has not been identified. There is also limited report on the modifications of fillers and fibers for enhancement of mechanical, physical (water absorption, density, etc.), thermal, and electrical properties (conductivity, dielectric properties, etc.) of LDPE matrix. Chemical recycling (pyrolysis) had been a major technology for waste or post-consumer LDPE to save the environment, but not cost-effective; emissions of some constituents and required additives or modifiers (catalysts) for considerable yields of the products in many applications. Incorporation of natural zeolite, clinoptilolite ($(K_2,Na_2,Ca)Al_6Si_3o_{72}$), improved the strength of the filled composites, rheological behavior, thermal, compatibility of the individual polymeric components, morphology, and texture of the moldings from recycled polyolefins which strongly depends on the type of zeolite, size and shape, branching, dimensions, and types of polymeric chains.

Chemical materials have been used as catalysts in the pyrolysis of plastics to obtain liquid products with higher yield and selectivity. Hence, numerous experiments were performed to find out the best catalyst to produce the most desirable products, taking the economic factor into consideration. Pyrolysis of plastic waste to fuel involves many limitations that prohibit the industrial plastic recycling process including the difficulty in modifying it from batch process to continuous process. In industrial process, plastic waste is fed into the reactor directly through hopper for melting in pyrolysis reactor with high melting point (300 °C and above, depending on the types of plastic). Therefore, any temperature lower than its meting point may result to solidification of the plastics in the process pipelines, hence causing blockage of the pipelines.

## Modifications and Properties of Recycling of Virgin and Waste PVC

The increase in commercial vehicles and road usage with construction resulted to increase in demand of bitumen for pavement and road construction. Yet, the durability of the bitumen depends on appropriate binder for enhancement of performance of bitumen. The use of little quantity of virgin thermoplastics provides a reasonable performance with bitumen but is uneconomical compared with only bitumen. The utilization of waste PVC for effective performance as bitumen binder in pavement and road construction products seems to be interesting because of its low cost and because it is one of the abundant thermoplastics that causes environmental threat. The applications of PVC have been reported to hinder and be not suitable for many applications because of incompatibility as a result of many factors. PVC possesses high melting points which hindered the mixing, and it is impractical to make any further attempts to incorporate it in some applications like bitumen road construction. Recycled LDPE/PVC blends have been modified using EPDM as effective toughening, compatibilizer, and dispersant agent in applications. Recyclability of PVC waste can be achieved mechanically without modifications or use of new plasticizer since the separation of other mixed plastics is possible through triboelectrostatic technology. The technology of triboelectrostatics depends on the ability of polymer to the electron loses or gains because electrons gains and charges negatively may be as a result of higher affinity of polymers, whereas loss of electrons and positively charge may be attributed to polymer with the lower affinity. Because of high electronegativity of chloride ions, it can mix with many polymers such as PET, PP, PS, and PE with enhanced properties. The use of wood fillers or fibers as natural modifiers have been reported to improve mechanical properties of recycled PCV rather the recyclability, and slightly reduction in mechanical (tensile, flexural, hardness and impact) and structural properties (i.e., decrease in molecular weight due to molecular chain scission caused by shear stress involved in reprocessing). The reduction in properties exists because of incompatibility or poor intermolecular interaction which can be modified by surface techniques.

## Modifications and Properties of Recycling of Virgin and Waste PET

Polyethylene terephthalate (PET) is a transparent semicrystalline, long-chain thermoplastic polyester which can be produced by a polymerization of terephthalic acid with ethylene glycol and remains the most used thermoplastics in many applications. It is characterized as easy to handle, durable, strong, thermally with low glass transition temperature, and chemically stable with low gas permeability. It exhibits brittle behavior, good mechanical properties, and dimensional stability as well as good gas and chemical resistance which resulted to its wide applications. Waste PET may be in bottles, foils, and cords from tire. Globally, the rate of generation of waste PET is about

20 million tonnes that amounted to about 15% which is alarming due to population growth, urbanization, standard of living, and cost of production, but the recycling rate of waste PET found to be 29.3% lower. The issue with the reuse of waste PET may be associated with size, content, mixing process, type of mixer, temperature, time profile during mixing process, and contaminations or additives like stabilizers and pigments. In bitumen asphalt modification for the road construction, the mixing process may be wet or dry process. The wet process involves blending of thermoplastics and bitumen in a mixer and then mixing of thermoplastic modified bitumen to aggregates, while the latter involves incorporation of thermoplastics to very hot aggregates prior to mixing with bitumen. Waste PET recycling employs dry process, and it can be modified to achieve better feasibility in terms of adhesion between the aggregates and binder, stability, and even mixing and minimizes the pore formation and moisture absorption. Appropriate recycling process conditions of waste PET make significant environmental and economy impacts through conservation of natural resources, environmental pollution, energy, and enhancement of engineering and physical properties of construction materials. An increase in recycled PET content caused a decrease in melt flow index or rheological properties of the aggregate. Recycled PET exhibits pseudo-plastic behavior, and it has been used to improve the rheological properties of asphalt as well as increased the viscosity and stiffness and enhanced the softening of stone mastic asphalt (SMA).

Incorporation of recycled PET with appropriate content and size increased the compressive, tensile, and flexural strength/s and ductility of concrete, creates lightweight aggregate of development of building materials, or decreases the bulk density of the composites, thereby helping polymer concrete in saving energy and minimizing the problem of solid waste posed by PET as well as other thermoplastics provided the impurities were removed prior to reprocessing.

Synthetic thermoplastics such as HDPE and acrylonitrile butadiene styrene (ABS) blend nano silicon (IV) oxide ($SiO_2$), and polylactic acid (PLA) can modify PET waste to improve its performance using the extrusion process based on a different mixing ratio. The use of virgin HDPE has been reported to improve rheological and mechanical properties when compared to waste PET using a less than 5% virgin HDPE. The mechanical properties of composites of recycled PET improved with increase in incorporated nano silicate ($SiO_2$) content blended with ABS. Modification of PET waste by the addition of small amounts of virgin PLA using melt mixing technology also shows reduction in viscosity of the composites with higher thermal sensitivity and mechanical properties compared to recycled PET. It should be noted that the performance of recycling of waste PET was hindered due to the presence of impurities, decomposition, and degradation of polymer chains. The modifications by compatibilizer like ethylene glycidyl methacrylate (EGMA) modified PE copolymer significantly improved the miscibility of recycled PET with PP, PE, and PS molecules, respectively, unlike linear low density polyethylene copolymer (LDPE). The use of natural materials to modify the properties of recycled PET such as fibers or fillers is not available. The efficacy and performance of recycled PET applications required optimum conditions of modified process, PET size and content, and additive or modifier content.

## Modifications and Properties of Recycling of Virgin and Waste Polypropylene

Due to favorable qualities of PP like density, versatility, photodegradation, and cheapness in cost of production, it is replacing many materials used for artifacts such as packaging products and automobile bumpers. The increasing rate of use of polypropylene coupled with inherent incompatibility of polyester and polyolefins seeks for improvement in the performance of PP in many applications. The improvement in PP performance has been achieved through modification techniques by incorporation of grafted maleic anhydride (PP-g-MAH), clay-based nano-fillers, inorganic nanoscale particles, and poly(trimethylene terephthalate) (PTT) blends using organically modified montmorillonite (Cloisite nanoclays) as compatibilizers for the purpose of improving compatibility, mechanical, crystallization, and melting behavior of PP composites. PTT is an aromatic polyester with combined properties of PET and poly(butylene terephthalate) (PBT). The factors that influence the properties of the PP composites are mix or blend ratio, crystallization temperature, compatibility process time, and size. There is loss of mechanical properties for composites of LDPE and HDPE modified with PP which is due to incompatibility of recycled PP/LDPE and PP/HDPE composites. The modification of recycled PP with HDPE reveals a partial compatibility which caused an improvement in tensile strength and elongation with the use of EPDM compatibilizer. The modification of recycled LDPE/PP with 1% montmorillonite nanoclay exhibits appreciable improvement in strength, physical properties, and stability of bitumen.

## Microstructural Behavior of Recycled Thermoplastic Matrix and its Modifications

The microstructural behavior in this content is limited to Fourier-transform infrared spectroscopy and scanning electron microscopy.

### Fourier-transform Infrared Spectroscopy

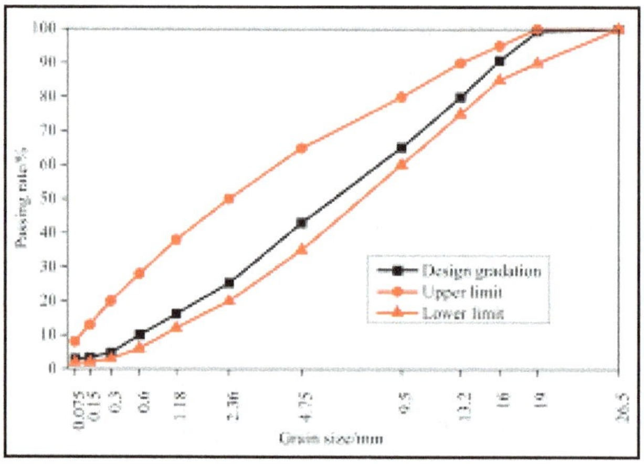


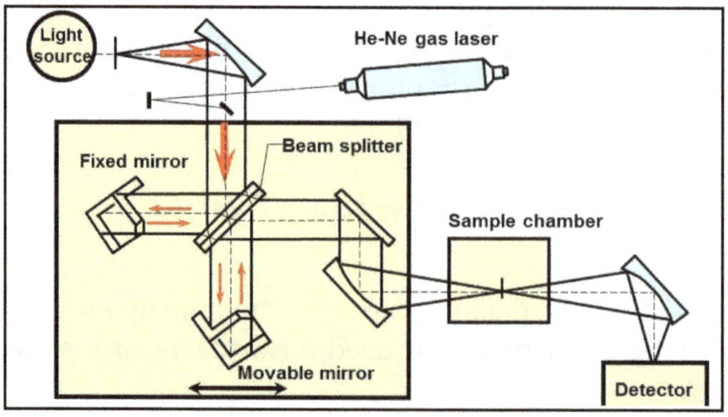

FTIR analysis of recycled thermoplastics exhibits no extra peaks for the blends, neither any shifts nor changes in the absorption bands of the carbonyl, hydroxyl, and carboxylic groups of HDPE, LDPE, PET, PVC, and PP resins which indicates the absence of any specific interaction, entanglement, or chemical reaction between the polymers and modifiers. In the case of modification of recycled thermoplastics using untreated natural fiber, there exists a shift or change in the absorption peaks of the carbonyl, hydroxyl, and carboxylic groups of the fiber-reinforced recycled thermoplastics, thereby influencing the physical and mechanical properties of the matrix and interfacial between the fiber and HDPE as reported by researchers. This resulted in improved quality of the thermoplastic products. The shift, change, appearance, and disappearance of absorption peaks correspond to reaction of the functional groups. This functional group dictates chemical reaction between the polymers and modifier, resulted to change in absorption peak correlate change in strength and modulus of the thermoplastics.

## Scanning Electron Microscopy

The scanning electron microscopy depicts the morphology of virgin and recycled thermoplastics at fracture surfaces when stressed and characterized the ductile, toughness, stiffness, and brittle nature of HDPE, LDPE, PCV, PET, and PP without modification, but improvement in compatibility using EPDM compatibilizer has been reported. The improvement in rheological morphology does not indicate an improvement in compatibility as well as mechanical properties. Modification of recycled HDPE with treated natural fiber using NaOH, SLS, acetic anhydride, $CaCO_3$ filler, and zeolites as well as synthetic fibers is characterized with improvement in polymer dispersion, even distribution of fibers, interfacial adhesion, fiber tearing, micro-crack formation, modifier content and size, nature of the modifier, and reduction in void formation. This indicated the enhanced compatibility which corroborates the improvement in physical, mechanical, and thermal properties of the modified recycled thermoplastics and dictates its applications.

## Application of Recycled Thermoplastics

## Applications of Recycled HDPE

The application of the HDPE composites is a function of the favorable properties coupled with cost implication of the production, and it may be affected by additional modified agents such as fiber or filler, NaOH, acetic anhydride, zeolite, and sodium lauryl sulfate. The use of recycled HDPE composites has been reported for many applications such as packaging (food storage containers and bottles for milk jugs), banners, swimming pool installation, corrosion protection for steel pipelines, folding chairs and tables, electrical and plumbing boxes, plastic surgery (skeletal and facial reconstruction), modified asphalt for pavement and road construction, housewares, industrial wrapping and gas pipes, and storage sheds, enhancing the economic, health, and social values as well as minimizing environmental issues that might be posed by HDPE disposal.

## Applications of Recycled LDPE

Incorporation of recycled LDPE at concentrations ranging from 2 to 5% by mass of bitumen possesses consistent desirable properties for bitumen asphalt applications. Utilization of LDPE for production of liquid milk packaging, bread packaging and sandwich bags, housewares, toys, buckets, wire and cable jacketing, and carpet and use of recycled thermoplastics for encapsulation of hazardous, radioactive, and mixed waste disposal save the environment from economic, environmental, and health issues.

## Applications of Recycled PVC

Polyvinyl chloride waste has been used in plumbing pipes and fittings, but its utilization as a binder in bitumen applications has been found unsuccessful due to high melting points which hindered the mixing as a result of poor compatibility. PVC sheets have been reported to be employed for making food trays, cling film and blister packages, household appliances, packaging, construction, medicine such as human rehabilitation, electronics, automotive and aerospace components, and building floor applications.

## Applications of Recycled PET

Recycled PET could be used for making waterproof water and soft drink bottles, thermally stabilized films (e.g., capacitors, graphics, film base and recording tapes, etc.), electrical components, and textile products if properly modified. The use of recycled waste PET as a modifier in bitumen road and pavement construction is hindered by mixing ratio and processing conditions due to high melting point. It is widely used in making automobile part, electronics, food packaging, house ware, lighting product, power tools, sports tools, x-ray sheets, and photographic applications.

## Applications of Recycled PP

Recycled polypropylene can be used for packaging articles, automobile bumper, foams, bottle tops, carpets, and household components and in making straws and sweet wrappings, PP powder, and PP mulch at concentrations ranging from 2 to 5% by mass of bitumen consistently desirable for bitumen asphalt applications. The recycled PP is also applicable in 3D printing filament. An application of recycled PP is dependent on good compatibility with modified materials with synergistic effects.

## Paper Recycling

The recycling of paper is the process by which waste paper is turned into new paper products. It has a number of important benefits: It saves waste paper from occupying homes of people and producing methane as it breaks down. Because paper fibre contains carbon (originally absorbed by the tree from which it was produced), recycling keeps the carbon locked up for longer and out of the atmosphere. Around two thirds of all paper products in the US are now recovered and recycled, although it does not all become new paper. After repeated processing the fibres become too short for the production of new paper - this is why virgin fibre (from sustainably farmed trees) will be added to the pulp recipe.

There are three categories of paper that can be used as feedstocks for making *recycled paper*: mill broke, pre-consumer waste, and post-consumer waste. *Mill broke* is paper trimmings and other paper scrap from the manufacture of paper, and is recycled in a paper mill. *Pre-consumer waste* is a material which left the paper mill but was discarded before it was ready for consumer use. *Post-consumer* waste is material discarded after consumer use, such as old corrugated containers (OCC), old magazines, and newspapers. Paper suitable for recycling is called “scrap paper”, often used to produce moulded pulp packaging. The industrial process of removing printing ink from paper fibres of recycled paper to make deinked pulp is called deinking, an invention of the German jurist Justus Claproth.

Waste paper collected for recycling in Italy.

## Rationale for Recycling

Industrialized paper making has an effect on the environment both upstream (where raw materials are acquired and processed) and downstream (waste-disposal impacts).

Today, 40% of paper pulp is created from wood (in most modern mills only 9-16% of pulp is made from pulp logs; the rest comes from waste wood that was traditionally burnt). Paper production accounts for about 35% of felled trees, and represents 1.2% of the world's total economic output. Recycling one ton of newsprint saves about 1 ton of wood while recycling 1 ton of printing or copier paper saves slightly more than 2 tons of wood. This is because kraft pulping requires twice as much wood since it removes lignin to produce higher quality fibres than mechanical pulping processes. Relating tons of paper recycled to the number of trees not cut is meaningless, since tree size varies tremendously and is the major factor in how much paper can be made from how many trees. In addition, trees raised specifically for pulp production account for 16% of world pulp production, old growth forests 9% and second- and third- and more generation forests account for the balance. Most pulp mill operators practice reforestation to ensure a continuing supply of trees. The Programme for the Endorsement of Forest Certification (PEFC) and the Forest Stewardship Council (FSC) certify paper made from trees harvested according to guidelines meant to ensure good forestry practices.

## Energy

Energy consumption is reduced by recycling, although there is debate concerning the actual energy savings realized. The Energy Information Administration claims a 40% reduction in energy when paper is recycled versus paper made with unrecycled pulp, while the Bureau of International Recycling (BIR) claims a 64% reduction. Some calculations show that recycling one ton of newspaper saves about 4,000 kWh (14 GJ) of electricity, although this may be too high (see comments below on unrecycled pulp). This is enough electricity to power a 3-bedroom European house for an entire year, or enough energy to heat and air-condition the average North American home for almost six months. Recycling paper to make pulp actually consumes more fossil fuels than making new pulp via the kraft process; these mills generate most of their energy from burning waste wood (bark, roots, sawmill waste) and byproduct lignin (black liquor). Pulp mills producing new mechanical pulp use large amounts of energy; a very rough estimate of the electrical energy needed is 10 gigajoules per tonne of pulp (2500 kW·h per short ton).

## Landfill Use

About 35% of municipal solid waste (before recycling) in the United States by weight is paper and paper products. 42.4% of that is recycled.

## Water and Air Pollution

The United States Environmental Protection Agency (EPA) has found that recycling causes 35% less water pollution and 74% less air pollution than making virgin paper. Pulp mills can be sources of both air and water pollution, especially if they are producing bleached pulp. Modern mills produce considerably less pollution than those of a few decades ago. Recycling paper provides an alternative fibre for papermaking. Recycled pulp can be bleached with the same chemicals used to bleach virgin pulp, but hydrogen peroxide and sodium hydrosulfite are the most common bleaching agents. Recycled pulp, or paper made from it, is known as PCF (process chlorine free) if no chlorine-containing compounds were used in the recycling process. However, recycling mills may have polluting by-products like sludge. De-inking at Cross Pointe's Miami, Ohio mill results in sludge weighing 22% of the weight of wastepaper recycled.

## Process of Paper Recycling

## The Step by Step Process of Paper Recycling

1. Collection: Recyclers and paper merchants collect the paper materials from collection points such as trash bins, paper stores, paper scrap yards, and commercial outlets that generate paper waste. Paper is collected from the bin and deposited in the large recycling container along with the paper from the other recycle bins. After collection, they are then measured, graded for quality, and hauled to recycling paper mill facilities.

This is the first process in the paper recycling process. Once the paper is collected from the recycle bins it is taken to the recycling plant where the waste paper is sorted and separated into types and grades.

With the arrival at the paper mill recycling facility, the papers are further checked for quality (cleanliness and type of paper) and quantity as purchasing contracts are issued based on these checks. Checking of the paper quality is also used to determine whether the type of waste paper is accepted or rejected. There are recyclers that accept mixed grades of recovered paper while some only accept preferred quality of waste paper grades.

2. Sorting: Once accepted at the recycling facility, the papers are further sorted based on quantity and paper value by assessing the materials that were used to make the paper. In most cases, the papers are classified according to their surface treatment and structure.

For instance, the very thin lightweight paper materials like newspapers are put separately from the thick paper materials like the ones used as paper folders. Sorting is important since paper mills produce different grades of paper materials based on the materials being recovered.

3. Shredding and Pulping: Once sorting is finished, the next step involves shredding followed with pulping. Shredding is done to break down the paper materials into small bits. After the material is finely shredded to bits, it is mixed with water and chemicals to breakdown the paper fiber materials. It turns the paper materials into a slurry substance, a process termed as pulping.

During the process of pulping, a large amount of water is added to the waste paper to produce pulp. Once pulp is produced it is then passed through a series of screens to remove larger pieces of contaminants for e.g.: inks, staples, plastic film and glue. The pulp material is then mixed up with new pulp to help the slurry substance solidify and form a firmer end product. The clean paper pulp is then placed in the machine that uses centrifugal cleaning to spin more of the debris from the paper pulp.

4. Filtering, conterminal removal and De-Inking: The slurry substance is then taken through a comprehensive filtering process to get rid of all the non-fibrous foreign materials present or any impurities such as strings, tape or glue. The pulp further goes into a chamber where contaminants like plastics and metals staples are removed by use of a centrifuge-like process. Light materials such as plastics float on top while the heavy materials like metals fall to the bottom for elimination.

The next process, de-inking, involves putting the pulp in a floatation device made up of chemicals and air bubbles that takes away any form of dyes or ink to enhance purity and whiteness. Hydrogen peroxide may also be used to further bleach the pulp. This entire step is also called the cleaning process as it cleans the pulp over and over to ensure it is ready for the final processing stage.

5. Finishing for reuse: This is the final stage of paper recycling. The cleaned pulp is blended with new production materials after which, it is put to dry on a flat conveyor belt and heated cylindrical surfaces. As the pulp dries, it is passed through an automated machine that presses out excess water. By the time the pulp is solid, it is passed through steam heated cylinders that facilitates the formation of flattened long rolls of continuous sheets of paper.

The resultant paper sheets are then trimmed, rolled, and sent to various business outlets or manufactures that use paper to make their products. Newspaper printing, wrapping papers, printing papers, and blown-in cellulose insulators are a few examples of areas where recycled papers are used.

## Metal Recycling

Metals are essential, versatile and can be used in a number of ways. Metals can be used for industrial purposes such as manufacture of trucks, cars, airplanes, ships, and railways. They can also be used to manufacture domestic items such as cutlery, crockery and even in packaging. The good thing about metal recycling is that, metal can be recycled over and over without altering its properties. The most common recyclable metals include aluminum and steel. The other metals for example silver, copper, brass and gold are so valuable that they are rarely thrown away to be collected for recycling. Therefore, they do not create a waste disposal crisis or problem.

## The Metal Recycling Process

The metal recycling process is similar to the usual recycling process. The metals are first sorted on the basis or their properties. It is however important to have the basic understanding or knowledge about metals. This will help in recycling them and keeping a green environment. The metal recycling process follows the following steps.

1. Collection: This is the first and most important step in metal recycling. It simply involves collecting all materials that are made of metals. This process should be organized in such a way that there should be containers specifically designed to collect metals. Some people and business persons have established scrap yards whereby people are encouraged to collect different metals, take them there and be paid for what they have collected. Different metals cost differently at the yards. The scrap metal yards are used as collecting centres for the metals.

2. Sorting: Once the metals have been collected, the next important step is to sort the metals. This involves separating what can be recycled form what is non-recyclable. It is essential to point out that the quality of metal recycled is very important. A high quality recycled product or item can only be created if the original materials used in the recycling process are of good quality. This therefore, calls for strict quality checkup during the sorting process.

3. Processing: After sorting, the next step is to compact or squeeze the metal. All the recycle materials are squeezed and squashed using machines so that they do not occupy so much space in the conveyor belts.

4. Shredding: After the crushing and breaking of the metal, the shredding process starts. The metals are broken down into tiny pieces or sheets to allow further processing. The small pieces have large surface to volume ratio that can be melted using less energy as compared to when they are in large pieces of metal. Normally, steel is changed into steel blocks while on the other hand aluminum is converted into sheets.

5. Melting and Purification: Melting of the scrap metal takes place in a large furnace. Each metal is taken to a furnace that is specifically designed to melt that particular metal based on its specific properties. The melting process uses a considerable amount of energy. However, the energy required for melting recycled metal is less compared to energy required to make metal from its raw material. The furnace is heated to appropriate degrees capable of melting the particular metal. Melting can take minutes or hours depending on how big the furnace is, the volume of metal placed in the furnace and the heat degree of the furnace.

6. Purification: After the melting process is complete, the next step is the purification process. Metals are purified using different methods. Purification of metals is done to ensure that the final product is free of impurities and that it is of high quality. Electrolysis is one of the methods of purifying some metals. Other metals are simply passed under powerful magnetic systems that separate

metals from other recyclables. Today there are different purification methods depending on the type of metal.

7. Melting and Solidifying of the Metal: After the purification process, the molten metal is then carried by the conveyor belt to a cooling chamber where it is cooled and solidified. It is at this stage that the scrap metal is made into a solid metal that can be used again. Other chemicals are then added into the molten metal to make it acquire its density and other properties. It is at the cooling stage that different shapes and sizes of metals are made and designed.

8. Transportation of the Metal Bars: Once the bars have been designed and made, the final product is then packed depending on their sizes and shapes ready for transportation to different factories and to people who require the metal. Thereafter, the cycle begins again.

## Importance of Metal Recycling

Some people disregard the need for recycling solid waste especially metal. Some throw them away while others leave the metal to rot away. However, if you understood the numerous benefits of recycling these items, you will no longer throw them away or leave them to be consumed by rust and other weather elements. The following are some of the benefits of recycling metals:

1. Preservation of Natural Resources: This is one of the main reasons why people should recycle metal. Metal is excavated from the earth through a process called mining. Over time, the areas excavated tend to be depleted and the miners move to other areas looking for metals to mine. This means that in case this trend does not stop or is not controlled, most areas will have huge excavation holes. Therefore, in order to prevent this and preserve the environment, it is important to recycle metal.

2. Reduction of Emissions: Have you heard of global warming? This is a phenomenon brought about by the steady increase of carbon dioxide emissions in the atmosphere. Although it is hard to undo what has already been done to the atmosphere in terms of destroying the ozone layer and global warming, we can control or stop causing further damage. This can only be done by embracing recycling of metal and other solid waste.

3. Economic Development: Through recycling, the government can save a lot of money that could be spent on mining and processing of metals. The money saved can be channeled into other

development programs that can help improve the living standards of the people and the economy as a whole. Additionally, through the exportation of recycled metals, the government can generate more revenue, improve the socioeconomic status of its citizens and provide jobs to the unemployed.

4. Management of Energy Consumption: The recycling of metal reduces the growing demand for natural resources such as ore. When more and more metals are recycled, the demand for the natural metal is reduced. On top of that, less energy is required to process scrap metal than it is required to extract ore from the earth, purify it and later shape it into desired shapes and sizes. The preserved energy can be used for other purposes in the economy.

5. Saves You Money: This is one of the best yet most ignored reasons. If most people recycled metal and metal items on a regular basis, it would be less costly to produce metal in terms of energy and the ore required. This will mean that the actual cost of buying metal will reduce hence saving you some money.

## Recycling of Ferrous Metals

Ferrous metals are mainly composed of iron and have magnetic properties. Steel, an iron alloy containing carbon, is by far the most-recycled material in the world. The most commonly recycled items are scrap from industrial processes, end-of-life products such as containers, vehicles, appliances, industrial machinery and construction materials.

The use of scrap metal has become an integral part of the modern steelmaking industry, improving the industry's economic viability and reducing environmental impact. Compared to ore extraction, the use of secondary ferrous metals significantly reduces $CO_2$ emissions, energy and water consumption and air pollution. At the same time, the recycling of steel makes more efficient use of the earth's natural resources.

## Recycling Processes

In general, metal recycling is a pyramid industry with many small companies at the bottom feeding scrap to large multinationals at the top. Steel recycling involves some, or,all of the following steps:

- Sorting: Because magnets attract steel, this metal can be easily separated from other recyclables like paper in a recycling facility with magnetic belts. Different kinds of steel do not need to be separated.
- Shredding: Shredders incorporate rotating magnetic drums to extract iron and steel from the mixture of metals and other materials.
- Media separation: Further separation is achieved using electrical currents, high-pressure air flows and liquid floating systems. Other processes may be necessary in cases such as steel cans which have a protective layer of tin that must be removed and recycled separately.
- Shearing: Hydraulic machinery capable of exerting enormous pressure is used to cut thick heavy steel recovered from railways and ships. Other cutting techniques, such as the use of gas and plasma arch, are sometimes employed.

- Baling: Iron and steel products are compacted into large blocks to facilitate handling and transportation.

A shredder in operation.

Recovery of aluminium and high value metals from an aeroplane.

## Applications

Steel is ideal for recycling because it does not lose any of its inherent physical properties during the process, which can be repeated ad infinitum. Steel is 100% recyclable and therefore, recycled steel can be used for the same applications as steel produced from virgin material. Products that are made of recycled steel include:

- Construction materials for roads, railways, infrastructure and buildings.
- Electrical appliances.
- Cans and containers.
- Automobiles and other vehicles.
- Office supplies.
- Hardware: Bolts, nuts, screws, etc.

## Recycling of Non-ferrous Metals

The most commonly used non-ferrous metals are aluminium, copper, lead, zinc, nickel, titanium, cobalt, chromium and precious metals. Millions of tonnes of nonferrous scrap are recovered annually and used by smelters, refiners, ingot makers, foundries, and other manufacturers. Secondary materials are essential to the industry's survival because even new metals often require the combined use of recycled materials.

## Recycling Processes

The metal recycling industry has an efficient structure with numerous small companies purchasing scrap material and feeding this to highly effective larger international businesses.

Non-ferrous metal recycling involves some, or all of the following steps:

Feeding recovered copper wire into a conveyor belt for recycling.

Baled zinc prepared for transpot to a refinery.

- Sorting: In order to be recycled appropriately, different types of non-ferrous metals need to be separated from each other, as well as from other recyclables such as paper and plastic.
- Baling: Non-ferrous materials are compacted into large blocks to facilitate handling and transportation.
- Shearing: Hydraulic machinery capable of exerting enormous pressure is used to cut metals into manageable sizes.
- Media separation: Shredders incorporate rotating magnetic drums to separate non-ferrous from ferrous metals. Further separation is achieved using electrical currents, high-pressure air flow and liquid floating systems. Further processing may be needed.
- Melting: The recovered materials are melted down in a furnace, poured into casters and shaped into ingots. These ingots are either used in the foundry industry or they can be transformed into flat sheets and other wrought products such as tubing, which are then used to manufacture new products.

## Applications

All metals can be recycled with minimal or no loss of their original physical properties. They are such versatile materials that the possible applications for each metal and their combinations are endless.

## Aluminium

Aluminium has (great recycling potential) and is often re-used for the same application for which it was originally manufactured. Its strength, flexibility and light weight, make it ideal for:

- Building & construction: Window frames, building structures, roofs, etc.
- Transportation: Airplanes, trains, boats, cars and trucks. It is also used in smaller vehicles like bicycles, motorbikes and other mobility devices such as wheelchairs.
- Packaging: Aluminium is used mostly in the form of cans and foil.
- Electricity: Since 1945, aluminium has replaced copper in high-voltage transmission lines.
- Cooking and tableware.

## Copper

After silver, copper has the best electrical conductivity of all the elements. It is also very good thermal conductor and is readily alloyed with other metals such as lead, tin and zinc for foundry applications to produce, among other goods, products for the transmission of water such as valves.

Other common applications for recovered copper include:

- Electrical applications: Wires, circuits, switches and electromagnets.
- Piping: Plumbing fittings and also in refrigeration, air conditioning and water supply systems.
- Roofing and insulation.
- Household items: cookware, doorknobs, and cutlery.

## Lead

Most of recycled lead is used in batteries, but the are many other applications for this metal:

- Car batteries: Lead is still used extensively in the plates that work as electrodes.
- Colouring: Although less common today, it is used in ceramic and glass glazing.
- Radiation protection: Lead offers protection against X-rays.

## Zinc

Zinc is present in everyday life in the form of coins. However, it also has other important uses:

- Galvanisation: Zinc is commonly applied as a coating to protect iron and steel from corrosion in a process known as galvanisation.

- Batteries: As an anode component material in batteries.
- Brass: Created by alloying zinc and copper.

## Tin

Apart from precious metals, tin is one of the most expensive non-ferrous metals. Hence, recycling from secondary materials is very important.

Its applications are very varied:

- Cans: By covering steel sheet with a thin layer of tin one obtains tinplate, the raw material to make cans.
- Car production: Tin increases the resistance of the motor block, piston rings and clutch plates.
- Springs of any kind become tougher through the addition of tin.
- Glass: Tin oxide coatings of glass surfaces to make them more resistant.

## Recycling of Steelmaking Plant Wastes in Clay Bricks

There are basically two main routes for steel production; they are the integrated route and the semi-integrated route. At the first route, the reduction of ore, including iron ore, coal or charcoal, and fluxing agents, occurs in the blast furnace to generate pig iron, in which it is then refined to produce steel, while in the second route the steel is directly produced in electric arc furnace.

During the steelmaking at both routes and also at the reefing step, the main residues generated can be divided into categories such as slag, sludge, and dust. The mineralogical and chemical compositions of each of these wastes vary according to the raw materials employed in the process and to what they have as common elements, mostly composed of iron, carbon, calcium, magnesium, silicon, and manganese and still containing smaller proportions of aluminum, zinc, alkaline earth metals, and traces of other elements.

It is preferable that the valorization of the solid wastes is carried out on the steelmaking route itself; however, if this practice is not possible, it is necessary to look for other productive processes capable of incorporating this waste as raw material. In association with this practice, it is desired that these wastes, when hazardous or not inert, are inertized in order to reduce their toxicity. In this sense, it is verified that much of this waste is still disposed in landfills or sent for incineration; even if this constitutes an environmentally correct practice, it is considered the least noble practice according to solid waste management.

The ceramic industry for civil construction has gained prominence in recent years due to its potential for incorporation of solid waste generated in the most diverse industrial activities. This potential is based primarily on the characteristics of the raw material and the high production volume in this industrial sector.

The basic raw material for the bricks and roofing tiles manufacturing is clay, and due to the nature of its formation, it has natural variability in its chemical and mineralogical composition. This variation, coupled with the low technical performance required for its products, allows residues of different compositions to be incorporated into the ceramic clayey body. In addition to direct environmental benefit, technical improvements in processing or product quality may occur, and even energy savings can be generated when certain wastes are incorporated with combustibility characteristics. Another notorious benefit is the possibility of inertization, in the ceramic matrix, toxic metals contained in these residues. Researchers still emphasize the function of improving sustainable development by providing the economy of clay, which is a natural nonrenewable resource.

## Steelmaking Plant Wastes as Raw Materials for Clay Brick Production

The ceramic industry as an active agent for the improvement of sustainable development.

The industrial segment of red clay-based ceramics, also known as structural ceramics, is part of the nonmetallic mineral sector of the mineral processing industry, sharing the entire production chain of civil construction. The following products make up this segment, bricks, roofing tiles, blocks, and paving ceramic floors, among others.

Several researches have reported on the state of the art of using various wastes to red clay-based ceramic products, such as red mud of Bayer process, fluorescent lamp glass waste, granite waste, effluent sludge from paper industry, blast furnace sludge, and sugarcane bagasse ash, among others. This potential is based on basically two particularities of this industrial sector, the characteristics of the raw material and the high production volume.

Researchers elaborated a classification in order to organize different types of industrial solid wastes as to their main influence when incorporated to the clayey bodies for ceramic production. Through extensive bibliographical research, these authors categorized the residues into combustible residues, fly ash, flux residues, and plasticity-modifying residues.

More recently, Researchers presented a new classification for industrial solid waste aiming its application as raw material for ceramic production for civil construction. In this classification, the authors propose only three categories, being combustible residues, flux residues, and property-modifying residues. Thus, fly ash was excluded from the old classification and should now be classified as combustible waste, and the category of plasticity-modifying waste was renamed as property-modifying waste, since not only plasticity but also other properties could be altered by waste that did not fit the other categories of classification.

## Characteristics of Steelmaking Wastes and their Influence on Clay Brick Properties

The steel industry generates a huge variety of wastes in the course of its process, mainly in the case of integrated mills. In relation to the solid wastes generated, the transformation of iron ore into steel can be classified into slags, sludges, and dusts, which together represent between 2 and 4 tons per ton of steel produced. The composition of these wastes varies according to the source of generation but usually contains iron, carbon, calcium, magnesium, silicon, manganese, zinc, and lead. Besides that, some slag and sludge contain a notable amount of heavy metals, and their improper deposition can negatively impact the environment.

So, the disposals of these wastes in landfills are becoming less attractive which not only occupy plenty of land but also increase the costs. Therefore, it is desirable to identify productive cycles capable of using the steelmaking wastes as raw material. In this way, this will lead to a better evolution of sustainable development. The following will be analyzed as to how the incorporation of these wastes in clayey bodies for production of ceramics for civil construction affects the properties and microstructure of these products.

## Slag

The slag originated at the steelmaking refining process, denoted as steel slag (SS), may vary in chemical composition as a function of the raw materials used and the technology and equipments employed. Currently, this residue is recycled in cement production as well as a base for pavements and composition for aggregates.

The difficulty of using the steel slag as raw material for construction products is its capability to enlarge. This is a consequence of the presence of free CaO and MgO as well as the polymorphism of the dicalcium silicate combined with the oxidation and corrosion of the metallic iron.

In relation to the characterization of this waste, the particle sizes with no crush processing present a wide distribution range varying from 20 to 30 × $10^3$ μm and mean particle sizes of 7.860 μm. This is an undesirable feature for traditional ceramic production, in which it typically uses raw material with particles having sizes lower than 2.000 μm. Aiming its application in an industrial way, it would be necessary to comminute this residue before it is added to the ceramic clayey body.

Table presents the chemical composition and the loss on ignition (LoI) of the steel slag, which is predominantly composed of Ca, Fe, Mg, and Si. Ca is present in the form of silicates as well as carbonate. Mg occurs mainly as free oxide and Fe as reduced oxides such as wustite and magnetite. Si occurs as silica and also as complex silicates. The loss on ignition may be attributed to carbonate decomposition and also to the dehydration of hydroxides such as brucite ($Mg(OH)_2$). The chemical composition results also in identified traces of K, V, Ni, Cu, Zn, Nb, Rh, and I.

Table: Chemical composition of the SS (wt.%) and loss on ignition (LoI).

| CaO | $Fe_{total}$ | $SiO_2$ | MnO | $SO_3$ | $TiO_2$ | $Al_2O_3$ | $P_2O_5$ | MgO | LoI* |
|---|---|---|---|---|---|---|---|---|---|
| 45.10 | 23.62 | 10.29 | 5.40 | 0.54 | 0.34 | 2.90 | 1.81 | 10.33 | 12.96 |

*Loss on ignition.

The incorporations of 0 (C0SS), 5 (C5SS), 10 (C10SS), 20 (C20SS), and 30 (C30SS) wt.% of SS into clayey body, sieved at 20 mesh (0.840 μm), were evaluated by `Vieira and Monteiro (2010). The specimens were sintered at 700, 900, and 1000 °C at heating rate of 3 °C/min.

The extrusion prognostic is an important parameter to production of ceramics for civil construction, since this property is related to the workability and productivity of the pieces. In figure, it can be observed that the clayey body without waste incorporation is located within the limits of the acceptable and non-recommended regions, and in practice, the results indicate that the incorporation of SS did not practically present significant variation at the workability of the clay.

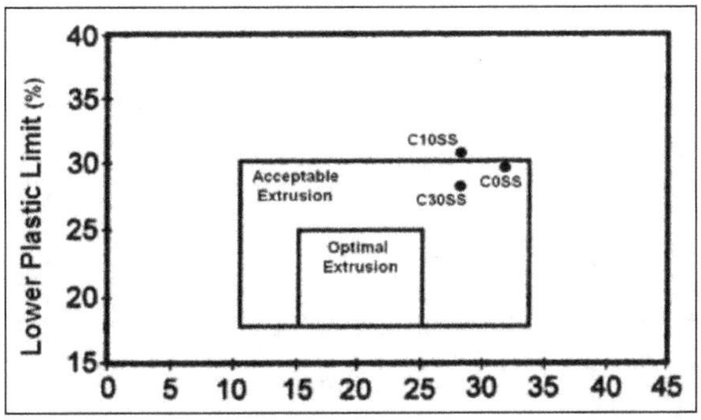


Extrusion prognostic through Atterberg limits.

The results of the technological properties evaluated are represented in figures. In relation to the influence of the steel slag incorporation, the incorporation of higher up to 10 wt. % impairs the water absorption, while adding up to 5% has improved this property. This slight improvement in water absorption may be due to the better packaging provided by the addition of the residue. It may be noted that the temperature elevation of 700–900 °C increased the water absorption; this can be explained by the combustion of coke fines at temperatures close to 800 °C. The linear shrinkage almost does not change with SS incorporation at sintering temperatures of 700 and 900 °C. From another point of view, at temperature of 1100 °C, the linear shrinkage significantly decreases with the SS incorporation. The flexural rupture strength is represented in figure, and the results suggest that the incorporationof 10% SS (CoSS) sintered at 900 °C is feasible for the reddish ceramic production since the values for this property are compatible with the products containing 0% of residue. At the lower temperatures investigated, 700 and 900 °C, the clayey body (with no SS incorporation) supports SS addition up to 10 wt.% without significant changes on its properties.

At temperature of 1.100 °C, the mechanical strength suddenly decreases with the amount of SS incorporation. At this temperature, the sintering mechanisms are highly intensified; in this sense, the clay minerals are extremely reactive during the sintering stages in temperatures higher than 950 °C due to their morphology, constitution, and particle size, promoting both solid-state and liquid-state sintering. Thus, the reduction in the content of clay minerals due to the incorporation of residue reduces the reactivity of the ceramic clayey body; therefore, there will be fewer points of contact formed during the sintering, and as a consequence, the mechanical strength of the product will be impaired. For this reason, the deleterious effect of the SS addition on the open porosity and mechanical strength of the clay is more pronounced.

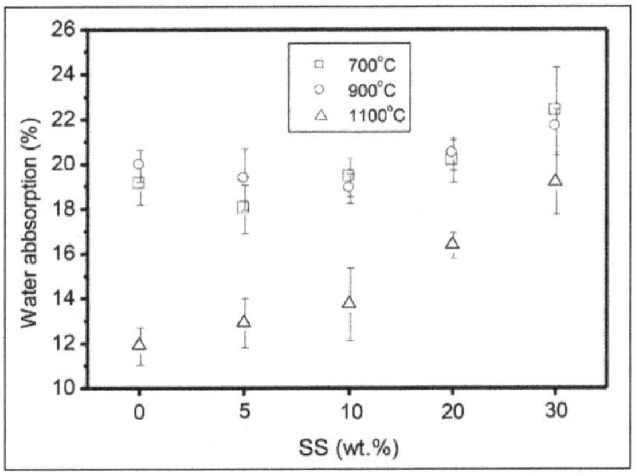


Water absorption of the clay as a function the amount of the SS incorporated.

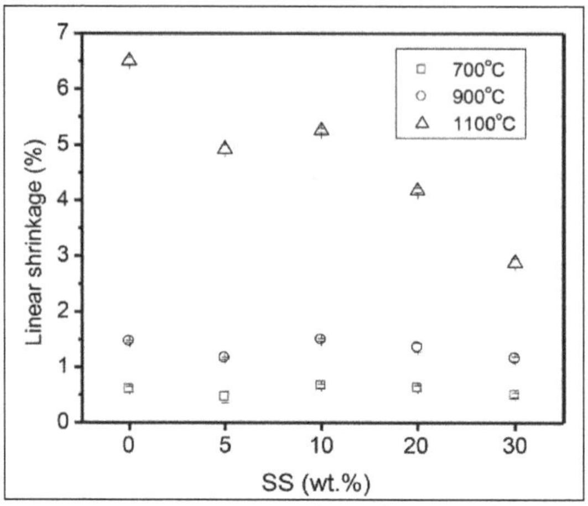


Linear shrinkage of the clay as a function the amount of the SS incorporated.

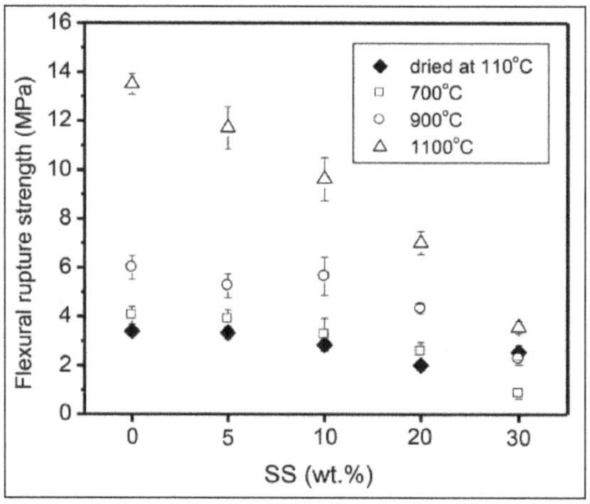


Flexural rupture strength of the clay as a function the amount of the SS incorporated.

The microstructure of the fracture region of the ceramics containing 0 and 30% of residue is shown in figures, respectively. At these micrographs, it may be noted that the two ceramics presented rougher texture. As expected, the ceramics sintered at higher temperatures produced samples with finer texture, but the cracks and voids are still present. It can be noted that the C30SS ceramic has typical regions in which the materials were plucked out at the moment of the mechanical resistance tests.

In this context, it is possible to predict that the SS particles do not react with the clay compounds. The insufficient level of chemical bonding leads the SS particles to stay loose from matrix and therefore generates defects such as voids and cracks in the product.

SEM micrographs of the fractured region of the clay (a) fired at 700 °C; (b) fired at 1100 °C.

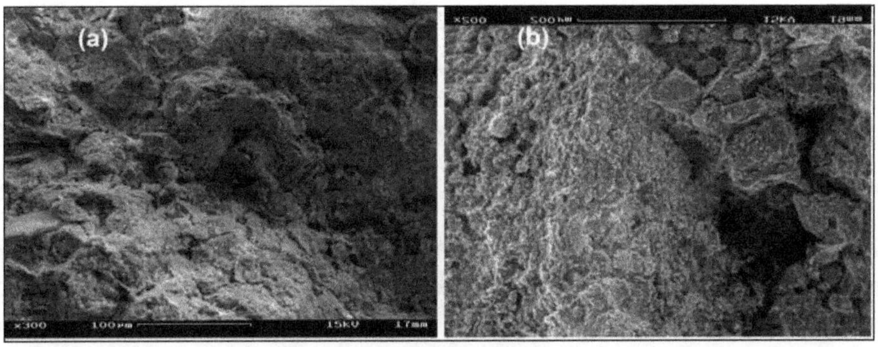


SEM micrographs of the fractured region of the clay with 30% of SS incorporated (a) fired at 700 °C; (b) fired at 1100 °C.

## Sludge

As a result of the gas treatment, sludge from wet dedusting systems, which have a high iron content with the order of 75 to 80%, is obtained from the implantation of the dedusting systems in the steelworks. Companies that have not developed ways to recycle this waste end up destining them to the landfills, excessively increasing the cost of production, since its volume generation is considerable, reaching values of the order of 36 $kg.t^{-1}$ of steel produced. One of the main problems for the recycling of steel slurry in the steelmaking process itself is its high moisture content.

The steel sludge has a particle size distribution that the equivalent spherical diameter varies from 1 to 800 µm and mean particle sizes with 35.6 µm. For reddish ceramic production, this is an appropriate characteristic, in which these industries use materials with particles having sizes lower than 2000 µm.

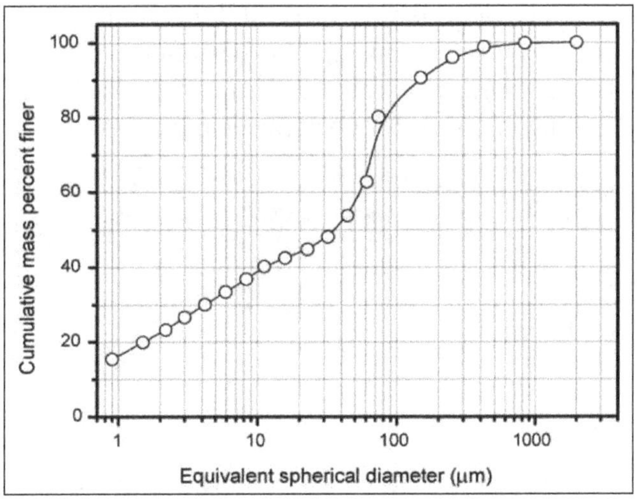


Steel sludge particle size distribution.

In relation to the chemical composition, the steel sludge is composed mainly by Fe, with 74.03%. The CaO comes from calcite introduced in the process to convert pig iron into steel. The small MgO content is associated with the desulfurizing agent also used in the steel process, which also contributes with CaO. The content of $SiO_2$ can be derived from the oxidation of silicon present in pig iron, while the ZnO content is associated with the use of galvanized scrap.

Table: Steel sludge chemical composition (wt.%).

| $SiO_2$ | $Al_2O_3$ | $Fe_{total}$ | $TiO_2$ | CaO | MgO | $K_2O$ | $Na_2O$ | ZnO | LoI* |
|---|---|---|---|---|---|---|---|---|---|
| 1.86 | 0.12 | 74.03 | 0.90 | 16.50 | 3.42 | 0.16 | 0.33 | 2.78 | — |

*Loss on ignition.

Figure shows the fracture surface micrographs of the ceramic with 5% of waste incorporated. A very rough microstructure is observed with the presence of well-distributed porous particles. These particles are associated with fine steel sludge and are not adhered to the ceramic matrix. This may lead to a reduction in the mechanical strength of products incorporating higher residue content.

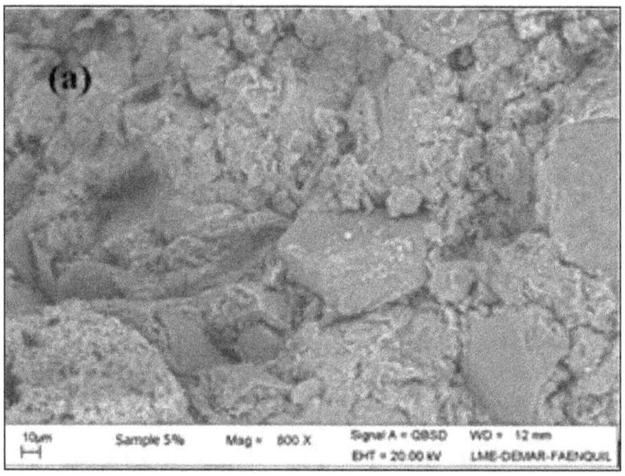


Micrograph obtained by SEM of the ceramic incorporated with 5% of steel sludge and sintered at 900 °C in magnification of 800×.

Figure shows the fracture surface micrographs of the clay with 20% of waste incorporated. It is possible to observe that the steel sludge is relatively well distributed in the clay matrix. However, with the increase of its quantity, more regions appear with fails. In addition, increasing the waste amount, which exhibits an inert behavior during firing, reduces the amount of clay, the main material responsible for the consolidation of the ceramic particles.

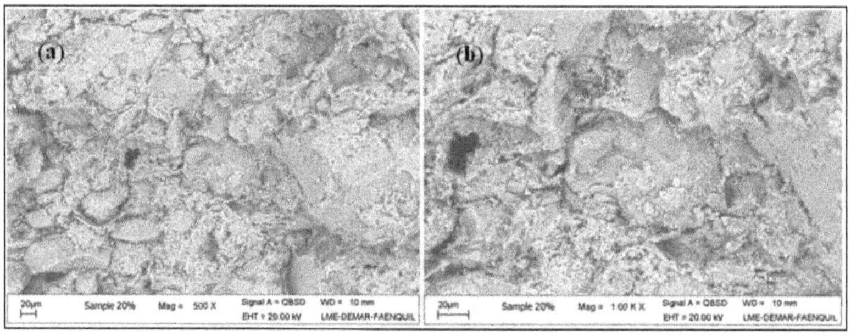


Micrograph obtained by SEM of the ceramic incorporated with 20% of steel sludge and sintered at 900 °C in magnification of 500× (a) and 1000× (b).

In table the leaching and solubilization results are presented, indicating the values obtained and the acceptable limit, according to the Brazilian legislation of potentially toxic metals of the formulation containing 5% of steel sludge. It is possible to observe that in the leached standard parameters, Ba, Cd, Cr, and Pb are present in quantities well below the limit values. Already solubilized, Al is present in a quantity nine times higher than the limit acceptable by norm. Al present in clay is one of its natural constituents and therefore apparently lies in equilibrium in nature. In this way its content above the stipulated limit by norm is questionable from the point of view of environmental problems. Other elements such as Cd, Cu, Fe, Mn, and Pb are in very small concentrations, being below the detection limit of the equipment.

Table: Potentially toxic metals in the leaching and solubilization extracts of the ceramic with 5% of incorporated steel sludge.

| Element | Solubilized (mg/L) | Limits (mg/L) | Leached (mg/L) | Limits (mg/L) |
|---|---|---|---|---|
| Al | 1.8 | 0.2 | 19 | Not required |
| Ba | 0.06 | 0.7 | 0.7 | 7.0 |
| Cd | <0.003 | 0.005 | 0.04 | 0.5 |
| Cr (total) | <0.02 | 0.05 | <0.05 | 5.0 |
| Cu | 0.004 | 2.0 | 0.13 | Not required |
| Fe | 0.02 | 0.3 | 0.22 | Not required |
| Mn | 0.02 | 0.1 | 0.9 | Not required |
| Na | 10 | 200 | — | Not required |
| Pb | <0.06 | 0.1 | 0.14 | 1.0 |
| Zn | 0.02 | 5.0 | 1.3 | Not required |

To evaluate the effect of steel sludge incorporation on the properties of ceramics for civil construction, Researchers prepared specimens with additions of 0, 5, 10, and 20 wt.% of fine steel sludge into a kaolinitic clay from Brazil.

It can be seen in figure the dried bulk density and the sintered bulk density of the formulations. One should notice that steel sludge waste (SSW) additions tend to increase the dried bulk density, mainly for the 20 wt.% addition. This behavior is probably due to the higher real density of the SSW particles compared with the clay particles and also the coarser particles of the SSW as compared with the clay, which may favor the packaging.

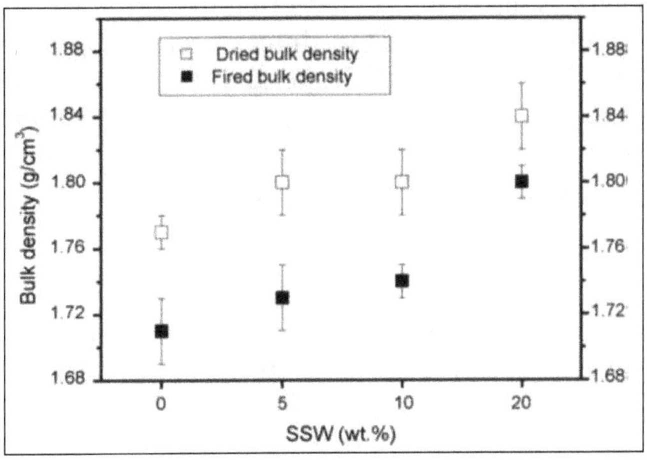

Dried and sintered bulk density of formulations containing steel sludge waste.

In figure, the results of the analyzed properties of the sintered ceramics as function of residue incorporation are represented. An interesting result can be noted for water absorption;

considering the error bars, this property is little affected by the residue amount increasing. Therefore, as the open porosity of the ceramics containing residue is practically unchanged, it is correct to affirm that the partial substitution of the clay by the residue aided in the reduction of the loss on ignition. It is known that the loss on fire caused by clay is due to dehydroxylation of kaolinite to form metakaolinite; in this way, this mechanism is responsible for leaving porosity in the ceramic. When the clay is replaced by the residue, although it is reducing the content of clay minerals which are the most reactive particles and contribute most to the consolidation mechanisms during the sintering, it is also reducing the loss to fire; for this reason, this property is little changed.

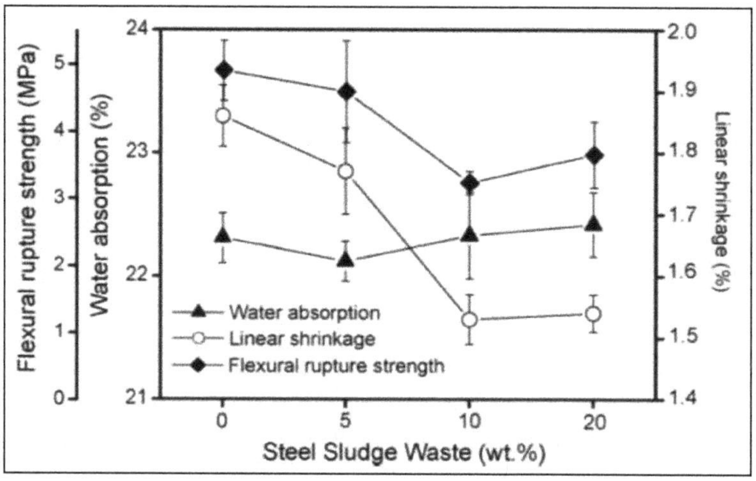


Sintered technical properties of formulations containing steel sludge waste.

This same explanation helps in the understanding of the linear shrinkage behavior. The loss on fire reduction with the SSW incorporation consequently causes the reduction of the firing linear shrinkage. This behavior assists in the ceramic production favoring the better control of the dimensions of the products, besides, of course, to avoid defects during the drying and sintering stages.

The flexural rupture strength is also represented in figure. Although the residue incorporation does not greatly affect the open porosity of the sintered ceramics, the mechanical strength is a property that is highly dependent on the particle consolidation during the sintering step. As seen, the mineralogical composition of this residue does not favor the sintering mechanism in these temperatures evaluated. In this way, additions of up to 5% by weight of waste are acceptable so that there is no too much reduction at the product resistance.

## Dust

Dusts are obtained in the dry cleaning systems of process gases from the sintering stage of an integrated steelmaking plant, here denoted as PMW (particulate material waste). Table shows the chemical composition of the PMW, which is predominantly formed by iron compounds such as hematite and magnetite. CaO and $SiO_2$ also present relatively high amounts. The loss on ignition is predominantly associated with the combustion of coke fines and the decomposition of both calcite and dolomite. The amount of 2.28% of $SO_3$ suggests the presence of sulfates, possible of Ca, such as gypsum.

Table: Chemical composition of the particulate material waste (wt.%).

| $Fe_{total}$ | $SiO_2$ | MnO | $SO_3$ | $Al_2O_3$ | $K_2O$ | MgO | CaO | ZnO | C | LoI* |
|---|---|---|---|---|---|---|---|---|---|---|
| 70.70 | 6.82 | 0.82 | 2.28 | 1.20 | 1.81 | 10.33 | 9.70 | 0.72 | 3.30 | 10.70 |

* Loss on ignition.

In figure, the particle size distributions of the PMW with no crush processing are represented. It should be emphasized that this particle size distribution prevents its reuse in the process to which it originated. However, for reddish ceramic industry, this characteristic is ideal, since this residue can be used directly without undergoing any other crush processing.

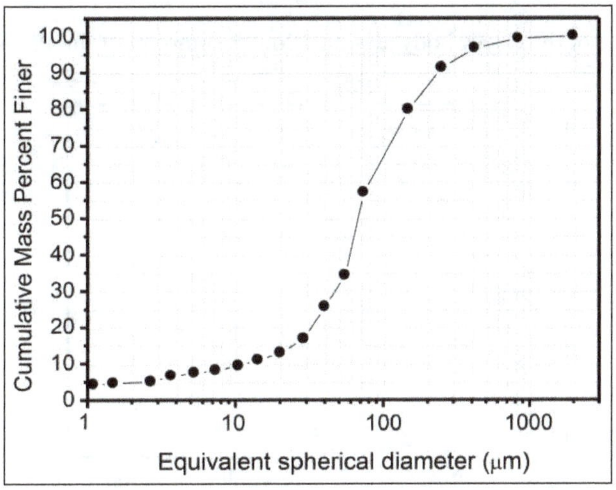


Particle size distribution of the PMW.

Figures show the graphs corresponding to the results of the technological properties after firing. A remarkable behavior can be noted with the water absorption reduction due to the residue content increasing, principally at 1050 °C. Although the ceramic clayey body was composed of highly refractory clay, the incorporation of 20% of the residue allowed the brick production at sintering temperature of 750 °C and tiles at 900 °C.

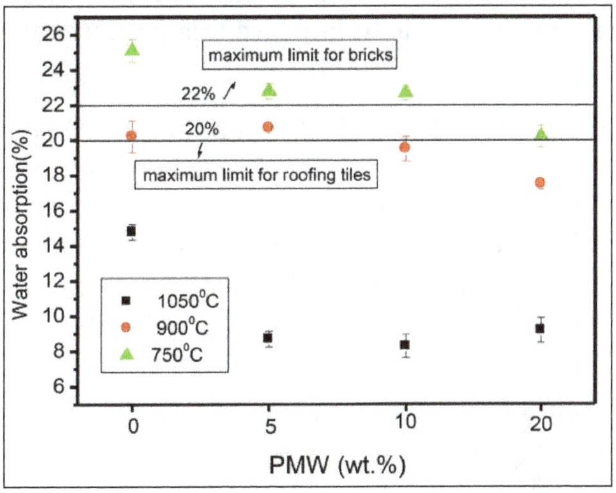


Water absorption of the clayey body as a function the amount of the PMW incorporated.

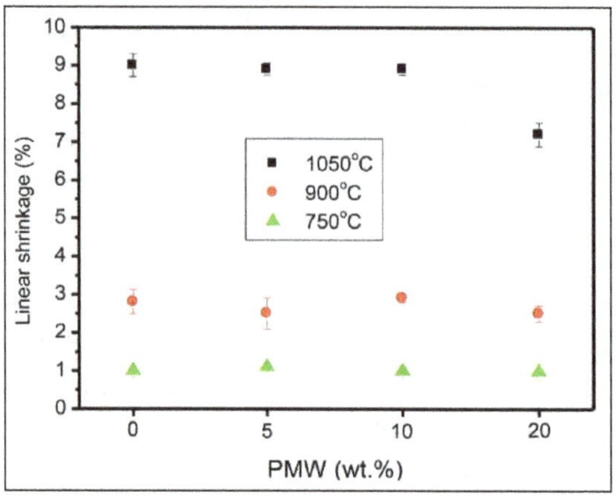


Linear shrinkage of the clayey body as a function the amount of the PMW incorporated.

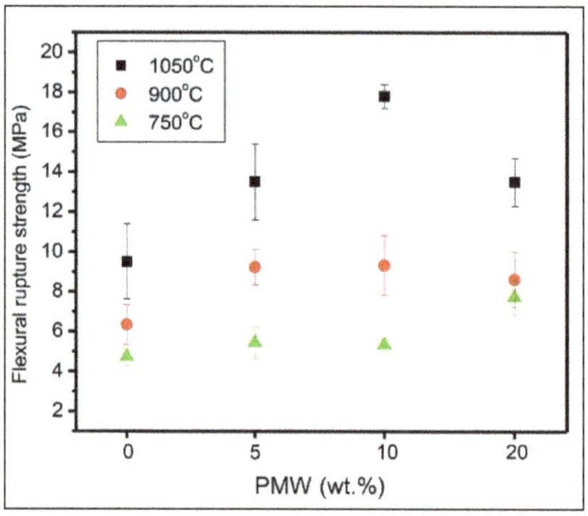


Flexural rupture strength of the clayey body as a function the amount of the PMW incorporated.

The firing linear shrinkage, represented at figure, shows that the PMW addition doesn't contribute to the increasing of this property; it is therefore a residue capable of assisting in the dimensional control of the products.

In figure, the mechanical resistance of the ceramics with PMW addition is represented. This residue with content lower than 20% was able to improve this property at all sintering temperatures investigated.

The mineralogical constitution of the PMW, which contains high amount of iron compounds, suggests an inert behavior during the sintering stage of red clay-based ceramics. Therefore, the explanation of how the residue may have acted to improve the properties may be based on dry bulk density. Probably the combination of these two raw materials helps in the better packaging at conformation stage favoring greater points of contact between the grains. Thus, there will be less void spaces which are responsible for the water absorption reduction and mechanical resistance increasing.

Figure represents the SEM micrograph of ceramic with 0% of residue content and fired at 1050 °C. The points located at this figure are derived from EDS analysis. The micrograph shows the intrinsic porosity probably related with void spaces left after the conformation step and consolidated as pores after sintering. The presence of pores can be associated with the dehydroxylation of some hydroxides and clay minerals and also with the different expansion coefficients among the ceramic phases during the sintering stage. Due to the refractory nature of this clay, the sintering is impaired as well as the reduction of pores, event at temperatures around 1050 °C.

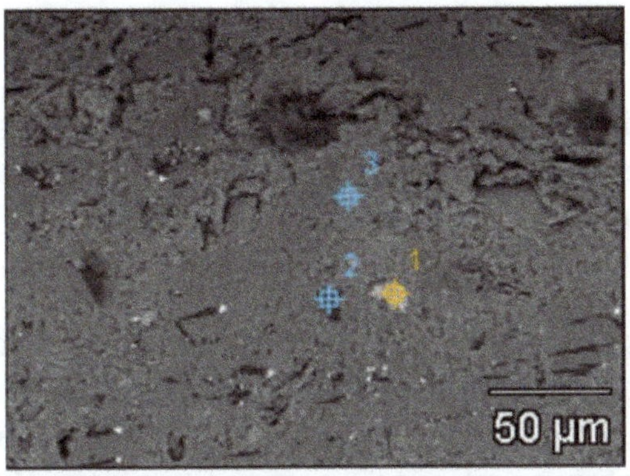


SEM micrograph of ceramic with 0 wt.% of steel dust incorporated and sintered at 1050 °C, including EDS spectra of selected points.

The EDS spectrum of point 1 indicates the existence of Zr and Si that can correspond to zircon silicate. Punctual regions 2 and 3 are basically composed of Si, Al, and Fe, i.e., the basic constituents of a kaolinitic clay. In this case, the Si and al predominantly constitute the aluminosilicate amorphous clayey matrix.

Figure shows two SEM micrographs of ceramic specimen with 20 wt.% of dust steel waste incorporated, sintered at 1050 °C. In this figure selected punctual regions, marked in the micrographs, were analyzed by EDS. In both micrographs of figure, defects associated with microcracks and pores can be observed.

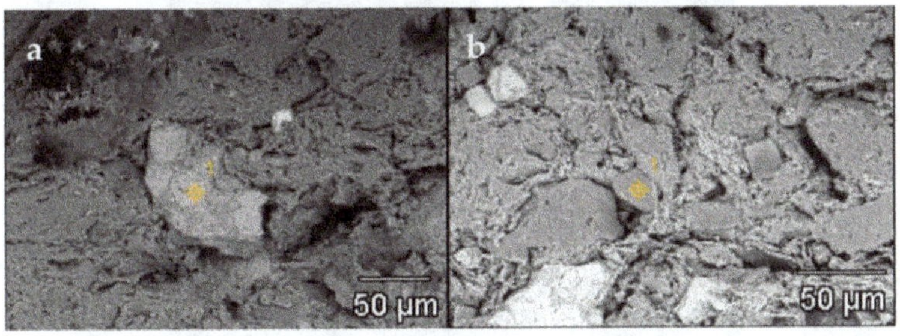


SEM micrographs of ceramic with 20 wt.% of steel dust incorporated and sintered at 1050 °C, including EDS spectra of selected points.

As already mentioned for the microstructure of ceramics with no dust waste addition, these defects are associated with the conformation step and further consolidation on sintering. It is important to notice in relation to figure that the main microstructural distinction between the ceramics with residue addition and those with no waste is the relatively larger particles apparently belonging to the waste. In fact, the EDS spectrum of the large particle in figure indicates the presence of Ca and Fe that are, as listed in table, characteristics of the PMW. Moreover, the microcracks that surround the particles indicate that this residue can lead to occasional defects in ceramics.

It is correct mentioning that, in spite of the defects observed, this waste acts as an inert material and assists on the plasticity control of the clayey body. Despite the inert character of the residue particles, the observed microcracks may be associated due their different expansion coefficient at sintering stage, more pronounced at 1050 °C. The particle analyzed in figure is relatively small and by its EDS spectrum, with Si and O, corresponds to a quartz particle, which already exists in the natural clay.

## Recycling of Polymeric Composite Materials

Composite materials are considered engineering materials that can replace non-ferrous or ferrous materials. Polymer composite materials have a large applicability in a different industries such as electrical engineering, electronics, building and civil engineering, rail, road and marine, aerospace technique and aeronautical, etc.

Composite materials consist of reinforcement material (glass fiber, carbon fiber, Kevlar, etc.) and a matrix (polyester resin, epoxy resin, and so on). Fiber glass is the most used reinforcing materials. They have many characteristics: high tensile strength, high chemical resistance, low cost. To obtain low price or to give high properties to a composite material we can include in a structure auxiliary materials, like: coupling agents, catalysts, pigments, accelerators and so on.

The storage of waste composite materials and the recovery of these, it's an important problem that we have nowadays. In Figure, it can be seen composite parts out of use and composite waste resulted from different production processes that occupy considerable spaces for storage. In time the accumulation of such materials can create serious problems to the manufacturing companies.

a.

b.

Composite materials waste stored.

Because the interest to find solutions for recovering or recycling is very low, the accumulation in time of composite materials waste is very significant. We can obtain a material rich in glass fiber by grinding the composite materials waste. Thus, it's obtaining a very valuable reinforcement that can be embedded in other materials or can be used for obtaining reinforced composite materials. A solution for recycling such composite materials has been to grind these materials, figure and create new composite products.

Ground glass fiber.

The recovery and recycling of polymeric composite materials has experienced an important concern in the last years. Researches dedicated to technologies for recycling composite materials were initiated and carried out by different authors.

Mixtures of concrete with sand and fiberglass waste are known. Waste composites can be used for concrete reinforcement or for a variety of construction materials. However, these materials from the known technical solutions point of view have higher density, lower mechanical properties, and the external factors like: UV radiation, moisture, sunlight, influence the degradation of these.

In order to make the ornamental plates used in the field of construction, a number of manufacturing processes are known which use sand mixtures with different binders: plaster, whitewash or cement. The obtained material as a dry mix or mortar is poured and pressed into a mold. After reinforcement of the material, the plate is extracted from the mold, after which time is left for stabilization, and then it can be used.

For the same purpose, for the production of alternative materials it is known the manufacturing process of reinforced mortars used in construction. These mortars include besides sand, whitewash, cement, gypsum and various reinforcement materials, such as: hemp fiber, glass fiber, etc.

The disadvantages of these processes are the high time of plates obtaining and their reduced mechanical characteristics. Other disadvantages of the plates obtained by these processes are the high density of the material and the degradation in time under the influence of external factors: humidity, sun, UV radiation.

Reinforced materials and manufacturing procedures have a significant influence on the quality, productivity and competitiveness of composite structures. The interface between matrix-reinforced materials plays an essential part in the mechanical behavior and fabrication of composite materials.

## Recycling of Composite Materials

The recycling of materials organic macromolecular surgery is more complicated than with traditional materials (glass, paper, metal), because there is an impressive variety of polymers, which in most cases are not compatible with each other, in the event of a global recycling.

For the recycling of polymeric materials there are several solutions:

- The separation of the constituents of mixtures in order to recycling of each individual component.
- The direct transformation of the mixture without prior sorting, in order to reduce the volume of waste.

From the view point of recycling, waste can be classified as:

- Manufacturing waste (10% of total waste) mainly formed of a single material. Because they are not contaminated (or less purified) with other materials, recycling is easier. Typically, these wastes are reintroduced into production lines.
- Waste easily separable. They consist of 1–2 or more polymers (mix macroscopic scale) otherwise contaminated materials (fillers). These materials are, at least theoretically separable.
- Microscopic mixtures or intimately connected (soldered, interpenetration). This is the case most difficult to treat because the separation of constituents is difficult or even impossible, requiring complicated operations. In this category fits very well with organic matrix composites. The most representative example is the waste from the automotive industry. In this case, the blend will find materials (resins) thermoplastic, polymer mix, fibers, fillers and multilayer composite materials.

The recycling after separation of mixtures is a much more interesting because you have to drive theoretically product with performance very close to the base polymeric materials. In practice, the properties of recycled materials approaching initial properties of the base material, unless methods are very effective waste sorting and waste have undergone significant degradation during operation.

Sorting of waste is done according to the basic polymeric material. The sequence of operations mainly comprises the following steps:

- Shredding: At this stage, the materials must be recovered shrinks size to be easily transported and handled.

  Separation of metals is well-established methods, obviating the mixture all existing metal fragments (e.g. electromagnetic methods).
- Shredding and spraying: This stage is complementary to the first mentioned. At this stage takes place and the washing waste. Choppers additional step is required for further processing easier.
- Washing and drying is intended to remove all impurities: In general washing is done with water and detergent, but depending on the nature of impurities at this stage can put complex problems. Flushing is required followed by spin drying to remove water.
- Separation using air or hydro cyclones and disposal are conventional methods for separating materials based on the difference in density. The process stream is brought into contact with a stream of air (water) in a cyclone. With these separation methods do not obtain high

degrees of purity. Moreover, this mineral phase (mineral fillers) often change the apparent density of the polymer, making it difficult to separate.

By materializing the proposed project creates prerequisites for achieving scientific and technological results, competitive at European level in order to increase the visibility of Romanian research, especially to subsequently transfer the results in socio-economic practice.

The resolving of proposed assignment will lead to the development of science-based knowledge of the manufacturing processes of parts of polymeric composite materials reinforced with biodegradable waste. The aim is to achieve a topic fundamental research, advanced to develop methods and technologies for recycling polymer composite materials and develop new recipes using biodegradable waste. It thus aims to improve the quality, productivity and competitiveness of industrial products. This is possible by using a multidisciplinary approach to research that brings together knowledge of chemistry, mathematics, physics, rheology and technological engineering.

## Proposed New Composite Material and Manufacturing Method for Ornamental Synthetic Plates

Both composite materials and waste composite materials resulting from production processes occupy important storage areas with an impact on the environment.

It shows the utilization of the glass fiber waste obtained by grinding the waste resulting from the manufacturing technological process of composite materials or removing them from their use and incorporating them in a product with applications in the field of industrial constructions, offering superior mechanical characteristics to the existing similar products.

## The Matrix

The mold, for obtaining the ornamental synthetic plates, is made of two separate modules, one of silicone rubber and one of fiber reinforced composite material.

Steps of mold forming.

Achieving the active part of the silicone rubber mold eliminates the need for additional separation planes, and the mold stiffness is ensured by reinforcing it with a fiber reinforced composite material. The use of matrix mold from the silicone elastomer eliminates additional separation planes reducing the cost of the mold.

The manufacturing process of the matrix, involves the following phases:

- Arranging the stones.
- Filling the joints with gypsum.
- Forming the outer frame.
- Applying the demulation layer, PVA (Polyvinyl alcohol).
- Preparation and application of the silicone rubber.
- Realizing the composite structural element for the silicone mold, fiber glass/polyester matrix.
- Demulation of the mold from fiber reinforced composite material.
- Demulation of the mold from silicone rubber.
- Trimming edges.

## Material

The process of ornamental synthetic plates manufacturing consists in a mixing, in a recipient, of calcium carbonate with the polyester matrix 5 minutes, casting these materials into a modular mold made from silicone and reinforced by a fiber-reinforced composite material and maintaining at room temperature for 20 minutes until the matrix gel point has been reached, mixing in another recipient for 10 minutes of waste glass fiber ground with a polyester matrix and 0–0.3 mm sand, and molding it in the die over the initial molded material, holding molds in the die approx. 2 hours at 60 °C, resulting an ornamental plate that is released from the mold after composite material polymerization. The material together with the unpolymerised matrix is deposited in a modular mold.

The sand was used as a low-cost reinforcement material in the form of particles with transparent aspect.

Sand grains.

The manufacturing process of synthetic decorative plates involves the following phases:

- Preparing the mold and applying the first layer release agent.
- Preparing the first mixture consisting of 60% polyester matrix and 40% Calcium Carbonate $CaCO_3$, mixed approx. 5 minutes in a recipient.
- Casting the first mixture so that it will cover more than 1–3 mm the height of the mold asperities and the maintenance until reaching the gel point, at room temperature.
- Preparing the consolidation mixture consisting of 30% sand of the 0–0.3 mm sort range, 30% grinded glass fiber waste and 40% polyester matrix and mixing it for 20 minutes.
- Casting the consolidation mixture over the first mixture until the mold filling and leveling the upper part.
- Transferring the mold with the composite mixture in a polymerization heat and maintaining in the mold at a temperature of 60 °C, about 2 hours, until the composite material polymerization.
- Mold release and obtaining the ornamental synthetic plates.

The composite material consists in obtaining a synthetic material composed of two component mixtures:

- The first mixture, which forms the surface layer that copy the mold and render the appearance of the synthetic plate, figure, consist of 60% polyester matrix and 40% Calcium Carbonate $CaCO_3$, mixed approx. 5 minutes, casted and maintained until the gel point was reached.
- The second mixture, of consolidation, consist of 40% polyester matrix, 30% sand of the 0–0.3 mm sort range, 30% grinded glass fiber waste, casted over the first mixture and maintaining in the mold until the polymerization, resulting a synthetic composite material reinforced with glass fiber having superior mechanical properties to similar materials used in construction.

The synthetic material consisting of two component mixtures.

The ornamental synthetic plate.

The percentages mentioned above represent the percentage of the total volume of the constituent materials.

The obtained material is a compact material with resistance at external agents, the process being easy to achieve. The composite material provides superior mechanical characteristics to traditional materials and can be used in other applications in the construction field such as reinforcing structures.

The following advantages are obtained:

- Composite material waste utilization, thus solving the significant problem of glass fiber waste.
- Enlarge the range of materials used in construction.
- Making a composite material having superior mechanical characteristics and low density with respect to traditional materials.
- Getting some plates with good look, imitating the natural stone, which can be colored in large quantity in the production process and can be easily mounted on facades and buildings.
- The technological simplicity of the process does not require substantial investment.
- The use of matrix mold from the silicone elastomer eliminates additional separation planes reducing the cost of the mold.
- Increasing the mechanical characteristics when using these materials at low temperatures.

## Electronic Waste Recycling and Disposal

Electronic waste, or e-waste, is said to be the fastest growing stream of hazardous waste in the world. E-waste is comprised of a variety of inputs including hazardous materials, potentially valuable and recyclable materials, and other inputs. E-waste follows a range of pathways after disposal, including formal and informal recycling, storage, and dumping, in both developed and less-developed country contexts.

This waste stream is comprised of any electronic items that a consumer or business intends to dispose of, or is no longer useful for its original purpose. E-waste has generated a considerable amount of public and political interest due to a confluence of factors, including: the exponential rise in the generation of e-waste, the potential value of recycling the waste in order to recover precious metals and other elements, and the environmental and human health risks associated with improperly storing, disposing of, and recycling e-waste. Some of the major responses to the rising generation of e-waste (and growing demand for secondary raw materials that it contains) have included the development of producer "take-back" legislation, technological innovations in recycling processes, and the formation of partnerships to facilitate the transfer of e-waste between the informal and formal recycling sectors.

E-waste is an incredibly complex waste stream, as it encompasses a wide range of items and the exact composition of many electronic components are considered to be trade secrets, meaning they are the confidential information of the manufacturer. Generally speaking, "modern electronics can contain up to 60 different elements; many are valuable, some are hazardous and some are both. The most complex mix of substances is usually present in the printed wiring boards (PWBs)" . To use a specific example, the material content of a mobile phone includes "over 40 elements in the periodic table including base metals like copper (Cu) and tin (Sn), special metals such as cobalt (Co), indium (In) and antimony (Sb), and precious metals including silver (Ag), gold (Au), and palladium (Pd)".

Electronics that had been used in industrial or business applications, such as medical equipment, have been recycled in the formal recycling industry for more than 40 years. These large items have frequently been exported within industrialized countries in the OECD to specialized facilities where they are processed for the purpose of extracting secondary raw materials. Consumer electronic waste from smaller items such as cell phones and televisions have not historically been profitable to recycle in countries with higher labor costs, since the quantity of recoverable valuable materials is relatively low. Hence, these items have typically either been disposed of, stored in consumers' homes, or exported (often illegally) to less developed countries such as China, India, Ghana and Nigeria, where they are recycled by informal recyclers using low-tech methods such as manual dismantling, open burning and acid leaching in order to recover gold, copper and other valuable metals. These methods generate subsistence livelihoods for workers but also result in significant hazards to human health and the environment as a result of the toxic materials that are also embedded in consumer electronics.

Generally speaking, the e-waste recycling process consists of five basic stages: collection, toxics removal, preprocessing, end processing and disposal. There are wide degrees of variation in how these stages are managed worldwide. For much of the global waste stream, e-waste may be collected informally via "waste pickers" or more formally through voluntary or mandatory producer "take-back" programs. In terms of consumer electronics, regions where e-waste is picked up by informal collectors have historically achieved significantly higher recycling rates than those where waste is dropped off through formal channels. After reaching the recycling site, dangerous components that require special treatment (e.g., batteries, Freon) are removed. The units are then separated into more homogenous groups based on material. This can be done manually, mechanically or a combination of both. Manual dismantling involves tools such as screwdrivers, hammers and labeled containers, while mechanical dismantling may involve conveyor belts, giant shredders and magnets.

Following the separation and dismantling phases, more homogenous groups of material (e.g., gold, copper, plastic, circuit boards) are then treated through a refining process: this can be accomplished chemically, with heat, or with metallurgical processes. This stage can be as high-tech as a giant smelter in Antwerp, Belgium or as low-tech as acid stripping in a backyard in Guiyu, China. Research has uncovered how sites will often compete for the waste by offering low-cost strategies, sometimes described as a "race to the bottom" process of increasingly lower standards and environmental protection. Finally, all of the components that cannot be sold or used as secondary raw materials are disposed of through means such as incineration or landfill.

The level of efficiency achieved through e-waste recycling depends upon the process that is followed, especially in the separation and dismantling phases. In dismantling electronics, manualized options are often much more effective than mechanized processes in gaining access to the best quality secondary raw materials. Mechanized take-back programs such as those in the E.U. do not even come close to the efficacy of the labor-intensive e-waste collection rates found in many African countries. Manual dismantling is also preferable to machine shredding, which damages and does not completely separate individual materials. For example, while 90% of the gold in discarded mobile phones can be recovered when manually dismantled, only 26% is recovered through mechanical shredding. However, these more labor-intensive options are not cost effective unless labor costs are extremely low.

## Secondary Raw Materials Recovered in Electronic Waste Recycling

E-waste contains components that have historically been valuable in significant quantities, when the dismantling costs have been low enough. Some of the applications and quantities extracted for different "important" or valuable elements within electronic devices are represented in table.

Table: A sample of valuable elements in electronic wastes.

| Element | Main applications | Total tons/year |
|---|---|---|
| Silver | Contacts, switches, solders | 6000 |
| Gold | Bonding wire, contacts, integrated circuits | 300 |
| Palladium | Multilayer capacitors, connectors | 33 |
| Platinum | Hard disk thermocouple, fuel cell | 13 |
| Ruthenium | Hard disk, plasma displays | 27 |
| Copper | Cable, wire connector | 4,500,000 |
| Tin | Solders | 90,000 |
| Antimony | Flame retardant; CRT glass | 65,000 |
| Cobalt | Rechargeable batteries | 11,000 |
| Bismuth | Solders, capacitor | 900 |
| Selenium | Electro-optic copier, solar cell | 240 |
| Indium | LCD glass, solder, semiconductor | 380 |

In addition to these metals, there is also another subset of elements—known as rare earth elements—which are crucial to the functioning of the newest electronics, particularly those with LED lighting and touch screen technologies. Rare earth elements are available in abundant quantities globally, but the process of their extraction can create widespread environmental problems, including radioactive contamination.

Table: Common uses of rare earth elements in electronic devices.

| Technology | Rare earths used |
|---|---|
| Electric and hybrid cars (NiMH battery). | Neodymium, praseodymium, dysprosium, terbium. |
| Computers (magnets in hard disk drive). | Neodymium, praseodymium, dysprosium, terbium. |
| Flat panel screens (glass coating to produce colors and brightness). | Yttrium, europium, terbium, gadolinium, praseodymium, cerium. |
| MRI machines (magnets). | Neodymium, praseodymium, dysprosium, terbium, yttrium, europium. |
| Smart phones (magnets and speakers). | Neodymium, praseodymium, dysprosium, terbium, yttrium, europium. |
| Other uses (including "chemicals, military weapons and delivery systems, and satellite systems". | Cerium, lanthanum, yttrium, neodymium, praseodymium, samarium, gadolinium. |

Recent technological developments, including improvements to the mechanization process as well as pilot projects that combine low-tech and mechanized methods, have been targeted to make e-waste recycling more profitable. Improvements to the mechanization process are fairly straightforward. On the one hand, revisions to shredding and sorting machines have improved the consistency and quality of the materials that are gathered at the preprocessing stage. In addition to this, newly mechanized methods are being developed to extract additional streams of secondary raw materials that were not previously recoverable. The major developments in this arena have been the invention of ways to extract various rare earth elements from electronics. State of the art facilities in Japan and France that can extract rare earths have recently become operational. Continued investment in technologies to recycle rare earths is seen as a strategic priority of industrialized countries, as these materials are essential for technologies related to communications, defense, and other state objectives, yet most mining for these materials takes place in China, a global power that has recently imposed quotas on the quantities that it is willing to sell for export. Concerns over the security and stability of the supply of rare earths have driven the development of new mechanized technologies to recover these materials from a wide range of e-waste inputs. Cost effective technologies for recovering secondary neodymium, dysprosium and praseodymium from e-waste are being further developed by U.S.- based recyclers and research institutes. Whether they are sited within the E.U., the U.S., or Japan, these newly operational recycling facilities will require a large quantity of e-waste inputs in order to be profitable. This challenge involves diverting a significant portion of e-waste from landfills, and from the informal recycling industry in less-developed countries.

## The Role of Extended Producer Responsibility in E-waste Recycling

Estimates of how much electronic waste is generated globally within a given year vary widely. These estimates are based on the quantity and volume of various electronic items that are purchased in a given year, with consideration to the anticipated life expectancy of that particular item. Surveys of recyclers on the volume of electronics collected can also be factored in, but it is important to note that a significant portion of consumer e-waste is either stored in consumers' homes or is mixed in with regular household waste and disposed of into landfills. In some instances, data on the amount of e-waste collected for recycling is available, such as in those regions that mandate producers to "take-back" consumers' unwanted electronics. Such mandates originated as part of the concept of extended producer responsibility (EPR), which holds that the manufacturers of products with hazardous components should bear the logistic and financial burden of recycling or disposing of their products in an environmentally responsible way.

While EPR legislation was initially opposed by manufacturers, the increased interest in the strategic importance and potential profitability of the secondary raw materials contained in e-waste (particularly the rare earth elements) has contributed to growing support for such legislation. Variations of EPR "takeback" laws have been put into effect across the globe, including in a number of U.S. states, across the European Union, and across many countries in Asia, Africa and Latin America. These laws signal a potential shift away from electronics recycling taking place primarily in the informal sector, and towards the growth of the formalized e-waste recycling industry.

With a growing number of EPR laws mandating manufacturers to take extended responsibility for the environmentally sound recycling of their products, there has been an increase in the number of pilot projects and public-private partnerships to collect and recycle electronics in ways that are efficient and cost-effective. Some of these projects entail the transport of e-wastes across national boundaries, and have fallen under the purview of the Basel Convention on the Control of the Transboundary Movements of Hazardous Wastes and Their Disposal (the Basel Convention). Over the years, The Basel Convention has convened technical working groups and conducted pilot projects, both of which have resulted in the development of technical guidelines for the handling and management of e-waste.

Under the purview of the United Nations University, additional pilot projects are being developed to facilitate a globalized e-waste recycling chain that involves labor-intensive dismantling and pre-processing in countries with lower labor costs (e.g., China, India, African countries), and high-tech end-processing in countries with more modern facilities (e.g., the EU countries). Major recycling corporations, electronics manufacturers, and government officials believe that such partnerships will insure a higher volume of input for large, high-tech smelters and provide access to the secondary raw materials that were previously "dumped" or otherwise retained within the global South countries where informal recycling currently takes place.

## Environmental and Human Health Hazards of Electronic Waste Recycling

The extent to which many of the other materials found in electronics are hazardous to human health and the environment is increasingly well-known. Electronics often contain toxic elements such as lead (Pb), cadmium (Cd), polychlorinated biphenyls (PCBs), polybrominated biphenyls (PBBs) and mercury (Hg) as well as other toxic components such as PVC and brominated flame retardants (BFRs).

Table: A sample of hazardous elements in an older-model PC.

| Element | Main applications | Weight (per 60 lb) | Dangers |
|---|---|---|---|
| Lead | Metal joining, radiation shield/ CRT, printed wiring board. | 3.8 | Human effects: neurological, blood, kidney damage. Brain damage and poisoning/death for children. Accumulates in the environment. |
| Mercury | Batteries, switches/housing, printed wiring board. | <0.1 | Human effects: long term brain damage and other neurological effects. Concentrates through the food chain. |
| Cadmium | Battery, blue/green phosphor emitter/housing, printed wiring board, CRT. | <0.01 | Human effects: acute and chronic damage to the kidneys. |
| Plastics (including those containing PVC and BFRs) | Casing, cable coating. | 22.99 | PVC effects: developmental toxin, reproductive toxin, endocrine disruptor. Carcinogenic when burned due to production of dioxins BFR effects: endocrine disruptor, neurotoxin, carcinogen (in humans and animals). |

Many of the health effects outlined in table have been documented in the town of Guiyu, China, where perhaps the greatest portion of the U.S.'s e-waste exports have been deposited historically. Here, almost 80% of children have respiratory problems, and they have an especially high risk of lead poisoning. Neurological, respiratory, digestive and bone problems are not uncommon among the workers and their families. In addition to these toxicological threats, which include long-term implications for both health and the environment, the threats posed by the recycling of e-waste are even greater when certain recycling methods are employed. For example, the informal recycling practice of burning plastic cables to retrieve the copper inside releases dioxins in the air via the burning PVC in the plastic. In more sophisticated operations (most typically found in Asian countries), a process of leaching printed circuit boards with acids (including nitric acid and hydrochloric acid) in order to maximize the amount of gold recovered can cause burns, respiratory and circulatory problems, pulmonary edema and death. The acid stripping leaves behind a toxic residue that oftentimes is disposed of in waterways where it can acidify water and destroy wildlife and vegetation. Heavy metal dust can travel to more populous areas and contaminate food supplies and greater populations.

There have also been studies on the occupational health and environmental risks associated with high-tech e-waste recycling, with experts noting that much more research needs to be done in this area in order to gain a more accurate assessment of these risks. Many of these studies are based on or informed by field research and experiments that measure concentrations of toxic chemicals in the workers, air, and environment around high-tech recycling facilities. There are indications in these studies that technologies such as the introduction of face masks and improved ventilation do decrease occupational exposures to a number of heavy metals and other hazardous chemicals.

In the U.S., a survey of 276 electronic waste recycling facilities was recently completed by the U.S. National Institutes of Occupational Safety and Health (NIOSH). It is especially relevant to note that this report finds that "most" of the responding facilities rely on manual dismantling, similar to the approach being applied in new and pilot facilities in less developed countries. It is also worth noting that a majority of the responding facilities were certified as environmentally sound either through the industry standard RiOS, the EPA standard R2 Solutions, or the activist standard e-Stewards. Hence, these facilities are likely to represent the "best case scenario" in electronics recycling. Overall, NIOSH concluded that "e-scrap recycling has the potential for a wide variety of occupational exposures particularly because of the use of manual processes". One of the primary concerns listed in the report is the potential for exposure to "metal dust" during the process of manual dismantling. Specifically, the report notes that it is unclear whether most facilities have installed proper filtration systems in order to remove metal dust from the air (since the majority of the facilities circulate air within the production area or rely on "natural ventilation"). The report also notes the use of compressed air for cleaning which can heighten exposure to metal dusts. While the initial NIOSH report says that acute exposure to heavy metals such as lead is unlikely, the report notes that "chronic lead poisoning, which is more likely at current occupational exposure levels, may not have symptoms or they may have nonspecific symptoms that may not be recognized as being associated with lead exposure".

Following the publication of the NIOSH report, additional studies of on-site occupational exposures in formal e-waste recycling facilities have been completed. Researchers reviewed 37 studies of the occupational hazards associated with formal e-waste recycling and concluded that, despite clear improvements to worker and environmental health when compared with informal recycling, "formal e-recycling workers and their families may experience unhealthful exposures to metals". The authors recommend further research "to reduce chemical exposures from formal e-waste recycling," along with the development of electronics components that are easier to safely disassemble, along with reducing the use of hazardous components in the manufacture of electronics. With e-waste now considered to be the fastest growing stream of hazardous waste in the world, there is an urgent imperative to implement solutions to reduce the risks associated with e-waste recycling.

## Hazardous Waste Storage

### Hazardous Waste Containers

Hazardous waste must be stored in containers (including lids) made of materials that are compatible with the waste. Hazardous waste containers must be in good condition and free of leaks or any residue on the outside of the container. Unacceptable containers include household detergent and food service containers. The best container for your hazardous waste is the original chemical container.

### Sealing Hazardous Waste Containers

Hazardous waste containers must be sealed to prevent leakage or spillage. Containers should be sealed with a screw-type lid or other appropriate device. Plastic wrap, aluminum foil, and other make-shift lids are unacceptable. A container holding hazardous waste must ALWAYS be closed during storage, except when it is necessary to add or remove waste.

## Labeling Hazardous Waste Containers

Hazardous waste containers must be labeled with hazardous chemical waste tags provided by VEHS as soon as the container is used to collect hazardous waste, regardless of whether the container is full. These tags require the laboratory to provide specific information including name, telephone number, building, room number, and exact contents of the container. The date should be filled out when the container is full. It is important to include as much information as is known about the contents of the hazardous waste container, including percentages and water content, to facilitate disposal. If a container is being used to collect hazardous waste intermittently, the tag should be filled out immediately upon use of the container and must be edited as more waste is added. A separate tag must accompany each individual hazardous waste container. Old labels that do not accurately describe the contents of the waste container (i.e., the original label for a toluene bottle now being used to store waste xylene) must be defaced.

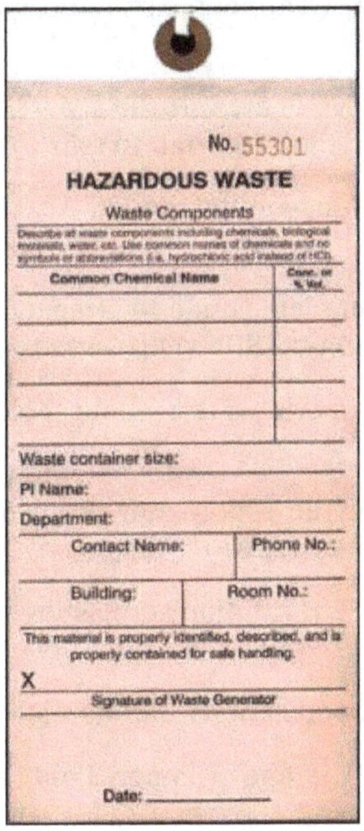


Pink chemical waste tag.

## Mixing Hazardous Wastes

Hazardous wastes should be kept separate whenever possible. Mixing a hazardous waste with a non-hazardous waste can increase the volume of hazardous waste for disposal. Mixing hazardous wastes with other hazardous wastes can increase disposal costs due to differences in disposal options for certain hazardous wastes. For instance, halogenated solvents such as methylene chloride and chloroform are more costly to dispose of than non-halogenated solvents such as hexane and xylene; therefore, halogenated solvent wastes should be kept separate from non-halogenated solvent wastes.

## Hazardous Waste Container Storage Area

You should designate an isolated portion of your laboratory as a hazardous waste storage area. Laboratories with multiple rooms may designate one hazardous waste accumulation area for all rooms as long as hazardous waste will not have to be transported in or across a hallway or through any area that is not controlled by the lab. Hazardous wastes must be stored with secondary containment so that spills cannot reach sink, hood, or floor drains. Incompatible hazardous wastes must be segregated to prevent reaction. Segregation methods include storing in separate cabinets, storing in separate hoods, or storing in separate secondary containment containers such as 5-gallon buckets or tubs. Please refer to your laboratory's Chemical Hygiene Plan or the VEHS Fact Sheet "Laboratory Activation Guide" for guidelines on segregating chemicals by hazard class. Post a copy of the "Hazardous Chemical Waste Accumulation Area Reference Sign" in your waste storage area to provide a quick reference for these procedures.

## References


- 5-methods-to-dispose-of-hazardous-waste: incinerationrecycling.com, Retrieved 17 August, 2019
- Earth-and-planetary-sciences/physical-treatment, earth-and-planetary-sciences: sciencedirect.com, Retrieved 09 June, 2019
- Baechler, Christian; DeVuono, Matthew; Pearce, Joshua M. (2013). "Distributed Recycling of Waste Polymer into RepRap Feedstock". Rapid Prototyping Journal. 19 (sur 2): 118–125. doi:10.1108/13552541311302978
- "Glass recycling information sheet". wasteonline.org.uk. Archived from the original on 18 November 2006. Retrieved 26 November 2006
- The Self-Sufficiency Handbook: A Complete Guide to Greener Living by Alan Bridgewater pg. 62—Skyhorse Publishing Inc., 2007 ISBN 1-60239-163-7, ISBN 978-1-60239-163-5
- Chemical-recycling-of-household-polymeric-wastes, household-hazardous-waste-management: intechopen.com, Retrieved 14 January, 2019
- Pet-bottle-recycling-for-sustainable-textiles, polyester-production-characterization-and-innovative-applications: intechopen.com, Retrieved 10 August, 2019
- Fleischman T. "Polymer additive could revolutionize plastics recycling". cornell.edu. Cornell University. Retrieved 23 February 2017
- Thermoplastic-recycling-properties-modifications-and-applications: intechopen.com, Retrieved 24 June, 2019
- Müller, Lothar (2014). White Magic: The Age of Paper. Polity Press. ISBN 978-0-7456-7253-3. Retrieved 28 December 2019

5

# Impacts of Hazardous Waste

Hazardous waste has many adverse impacts on the human health and environment. Air, soil, water, wildlife, etc. are some of the biotic and abiotic factors of the environment which are severely affected by the amounts of hazardous waste. The topics elaborated in this chapter will help in gaining a better perspective about the impacts of hazardous waste.

Nearly every activity leaves behind some kind of waste in the environment. Households create ordinary garbage. Cars, trucks, and buses emit exhaust gases while in operation. Industrial and manufacturing processes create solid and hazardous waste. Some wastes contain chemicals that are hazardous to people and the environment. Once these hazardous chemicals are present in the environment, people can become exposed to them. Exposure occurs when people have contact with a chemical, either directly or through another substance contaminated with a chemical.

The place where the chemical originates is called the source. Chemicals can enter the environment from many different sources such as landfills, incinerators, tanks, drums, or factories. Human exposure to hazardous chemicals can occur at the source or the chemical could move to a place where people can come into contact with it. Chemicals can move through air, soil, and water. They can also be on plants or animals, and can get into the air we breathe, the food we eat and the water we drink.

The different ways a person can come into contact with hazardous chemicals are called exposure pathways. There are three basic exposure pathways: inhalation, ingestion, and skin contact. Inhalation is breathing or inhaling into the lungs. Ingestion is taking something in by mouth. Skin contact occurs when something comes in direct contact with the skin. Ingestion can be a secondary exposure pathway after skin contact has occurred, if you put your hands in your mouth and transfer the chemical from your hands to your mouth.

The Environmental Protection Agency (EPA) has developed informational summaries on selected chemicals to describe how people might be exposed to these chemicals and how exposure to them might affect their health. The summaries also explain what happens to the chemicals in the environment, who regulates them, and whom to contact for additional information.

Some common ways a person may be exposed to hazardous chemicals include:

- Water: Exposure can occur when people drink contaminated groundwater or surface water, or accidentally ingest it while swimming or showering. Direct skin contact also is an exposure pathway that occurs during activities like swimming and showering. For more information about water and hazardous chemicals, view the MDNR, CDC and EPA websites.
- Soil, Sediment, or Dust: People can be exposed to hazardous chemicals in soil, sediment, or dust if they accidentally ingest it, breathe it in, or have direct skin contact. Children are highly susceptible to these exposure pathways. In their daily activities, children have a tendency to have frequent hand-to-mouth contact and introduce non-food items into their mouths.
- Air: Exposure can occur when people breathe in hazardous chemical vapors or air that is contaminated by hazardous chemicals or dust. More information is available about chemicals in the air at the websites of MDNR, CDC and EPA.
- Food: People can be exposed to hazardous chemicals through the food they eat. Food contamination can occur if the food has come into contact with hazardous chemicals. It can also occur further down the food chain such as through eating contaminated fish.

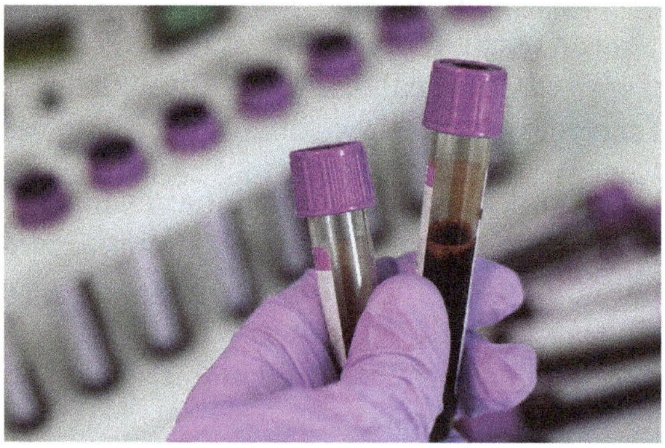

Exposure, the chemical getting into or on your body, has to occur to make you sick or cause adverse health effects, or have any affect on your health. If you are not exposed to the chemical, it cannot make you sick. Also, even if exposure has occurred, adverse health effects may not occur.

Adverse health effects are dependent on the factors of the exposure. Factors that play a part in whether or not adverse health effects may result from an exposure are:

- The type of chemical.
- The amount or dose (the amount or level of a chemical a person was exposed to).
- The duration (how long did exposure occur).
- The frequency (how many times the person was exposed).

Also, the occurrence of adverse health effects can depend on the way the chemical enters your body. Some chemicals rapidly absorb through skin, others not at all. Health effects also depend on the toxicity of the chemical that entered your body. Some chemicals are very toxic in small amounts; others are only toxic in large volumes.

Also, people respond to chemical exposure in different ways. Some people may be exposed to a chemical and not get sick. Other people may be more sensitive to chemicals and get sick more rapidly or have more severe reactions than others. Certain variables play a role in a person's susceptibility to exposure and adverse health effects such as age, gender, genetics, pregnancy or other health conditions.

For the fetus, child, and adolescent, the adverse health effects from exposure to chemicals can be much greater than for adults. The factors that affect their susceptibility include their stage of development and level of activity in the environment. The fetus is the most susceptible because their developing organs may be permanently damaged. Similarly, children, especially from one to six years of age, are also in a stage of rapid development. During this period, children may take more chemical into their bodies due to body chemistry, level of activity, and relative small body size. As children develop, chemicals introduced into the body can alter many processes essential for proper cell development. As a result, changes can, for example, cause organs within the body to be altered, impairing proper development to a mature organ. Adolescents may share in the risk due to their increased physical activity and curiosity of the world around them.

The human body has the ability to tolerate certain amounts of chemicals and the ability to excrete chemicals from the body. Once a person is exposed to a chemical, it may enter the blood stream, and eventually reach the liver. The liver attempts to detoxify harmful chemicals in the body by converting them to less toxic ones or ones that could be used by the body. The body naturally attempts to eliminate substances that are harmful or are not used. The kidneys filter substances out of the blood and excrete them in urine. Also, chemicals are removed from the body in feces, sweat and exhalation. However, the body may not be able to remove all the chemicals. The amount, type, and length of time you are exposed to harmful substances will determine if you are at risk for adverse health effects.

Therefore, the state and federal governments have determined health-based comparison values for contaminants that when exceeded, exposure may result in an excessive health risk. The chart below contains levels used by agencies to determine if doses of chemicals may be harmful to human health.

## Impacts of Hazardous Waste on Human Health

Hazardous waste carries environmental risks and also health risks for humans and wildlife. Some pollutants such as mercury can accumulate in human and animal tissue, thus compounding their effects. Hazardous waste is primarily generated by industry and businesses. Although regulations exist, contamination still occurs. In 2009, the U.S. Environmental Protection Agency (EPA) recorded 23 million cases of voluntary disclosure of pollution risks and opened 387 environmental criminal cases. As long as threats remain, health effects of hazardous waste will continue to occur.

### Cancer

American farmers apply more than 300 million pounds of pesticides to farmlands each year. Of the 27 most commonly used pesticides, the EPA has classified 15 of them as carcinogens or cancer-causing agents. Cancer has also been linked to air pollution from industry as well as in the home. Radon, for example, is a radioactive by-product of uranium decay. Uranium is found within the Earth's crust and is everywhere in the environment. Radon exposure is the second leading cause of lung cancer according to the National Cancer Institute.

### Respiratory Conditions

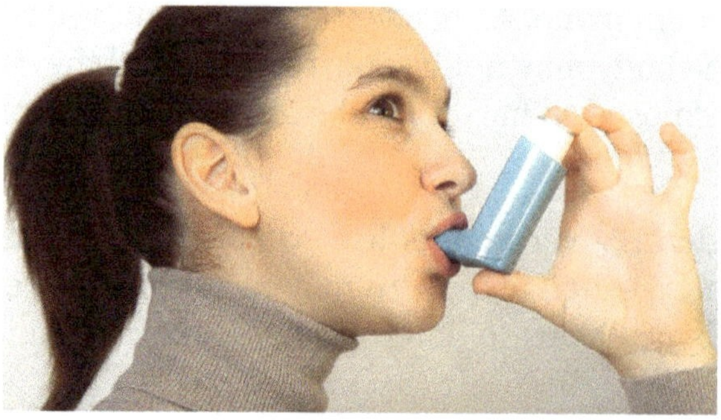

A direct link exists between air pollution and respiratory conditions such as asthma. Exposure to hazardous waste from emissions irritates the mucous membranes of your mouth and throat. A 2008 study published in the Annals of the New York Academy of Sciences found that individuals merely living near a hazardous waste site had an increased risk of developing respiratory diseases.

## Heart Disease

The risks of living near hazardous waste sites do not stop with increased risk for respiratory disease. A 2004 study published in the Archives of Environmental Health found an elevated risk of the development of congenital heart disease in the offspring of pregnant women living within one mile of a hazardous waste site. The threat is also more innocuous. Auto emissions also carry an increased risk of heart attack and stroke from thickening of arteries. Fossil-fuel emissions contain several toxins considered non-specific hazardous waste by the EPA. A non-specific hazardous waste is one without a readily identifiable source.

## Exposure Effects

The health effects from some types of hazardous waste may be temporary, with no link to other conditions having been determined. Xylene, for example, is one of the most widely used chemicals in the United States. It is an ingredient found in paints, solvents, and varnishes. Although not considered a carcinogen, exposure to the chemical causes dizziness and headaches. A person may also experience stomach discomfort. At high levels, xylene may cause unconsciousness and even death.

# Environmental Impact of Household Hazardous Waste

Improper disposal of HHW eventually leads to the presence of hazardous contaminants in the environment. All the facilities that are used to manage discarded HHW are in direct contact with environment media, air surface water, groundwater, and soil. These media are in constant contact with each other. As a result, when facilities cannot adequately break down hazardous compounds in HHW, the immediate environment is at risk.

The contaminants enter the water cycle via groundwater or lakes, rivers, and streams traveling through the cycle via different paths:

- Precipitation from the atmosphere.
- Percolation through the soil.
- Direct disposal from a wastewater treatment plant (WWTP) into a surface water body.
- Residents pouring liquids down the stormwater drain that empties into a lake.

In addition, toxic gases from HHW can be emitted into the air from the hazardous compounds that are used in producing them during controlled incineration or sometimes, uncontrolled fires.

## Landfills

Landfills can be the most economic way for waste management, especially in countries like Canada with large open spaces. However, poorly managed landfills have the potential of causing a number of environmental issues such as contamination of groundwater or aquifers or soil contamination. Modern landfills are not just holes in the ground to be packed with garbage. They can be considered as highly engineered contaminated systems. A modern landfill uses a number of technologies to ensure that the wastes are properly managed to avoid environmental pollution (e.g., ground water contamination, gas emission). Figure shows schematic of a modern landfill process. Advanced protective liners (both natural and manufactured) are typically used to isolate the waste and leachate from leaking into the surrounding ground or ground water. Single, composite, or double liners can be used depending on the nature of the waste materials being deposited. At minimum, a composite liner should be used for hazardous waste landfill facilities. However, landfills are not usually engineered to handle toxic compounds from HHW. Hazardous liquid waste can be transported from a landfill into the environment if there are no barriers. Leachate that has been contaminated with hazardous material (soluble or insoluble) may destroy synthetic liners and render existing barriers ineffective, and thus, the hazardous waste comes in contact with the soil. Its fate is determined by the characteristics of the soil such as porosity, geological factors, and the contaminant like viscosity. The contaminant may percolate downward and affect the groundwater or spread and contaminate surrounding area. Even if the leachate is collected, the treatment plants are not usually equipped to remove hazardous contaminants and end up releasing them into water bodies.

In addition, the conditions of the landfill such as the air and moisture content can affect the fate of hazardous contaminants such as the rate of degradation or violent reactions. For example, phthalic acid esters (PAEs) are used as plasticizers that are used in furniture, clothes, food packaging, etc., which are items that will invariably end up in the landfill. While readily degradable under aerobic conditions, those that are found in the landfill environment tend to retard biodegradability of PAEs. When the environmental impact of PAE in a landfill in China was studied, it was discovered the more complex congeners were found absorbed in deeper soils and in the groundwater.

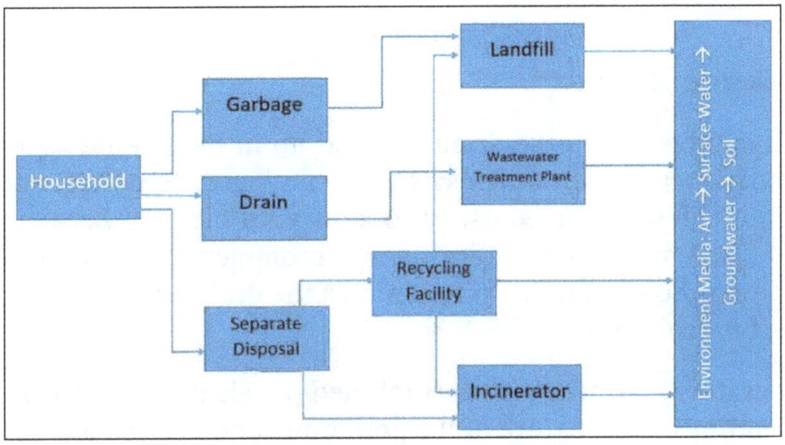


Improper disposal path of HHW from household to environment.

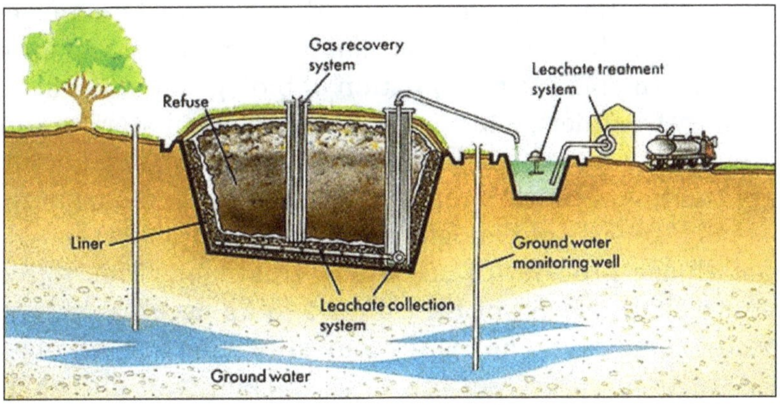


Schematic of modern landfill process.

Certain volatile organic compounds can be partially degraded and are readily absorbed by MSW in a landfill rather than volatilize. The moisture in the leachate enhances this process. Leachate-containing toxic compounds can be detoxified faster by recirculation within the landfill, which reduces the potential for leakage from the landfill liner. HHWs contribute volatile organic compounds (VOCs) to landfill gases such as benzene, methylene chloride, trichloroethylene, vinyl chloride, etc. VOCs from landfill gases contaminate off-site groundwater through migration.

## Incinerators

The quality of air emissions and ash residue is as a result of the fuel being incinerated. Incinerators usually have pollution control devices; however, some of the components that are found in HHW can pose a challenge to be captured. For example, mercury found in dry cell batteries, fluorescent light bulbs, and old paint can be converted to gaseous form and be emitted from the stack. Even the use of air treatment technologies can only remove 75–85%. Once it becomes in the atmosphere, mercury can be solubilized by rain and end up in water bodies. Other contaminants such as hydrogen sulfide and carbon monoxide that enter the atmosphere as gases may react with other compounds to become even more hazardous or remain in the atmosphere if stable, causing damage. Also, toxic metals have been found in the fly ash residue of incinerated MSW containing HHW. Damaging explosions have been reported due to a flammable liquid container being heated, which can lead to a few hours to few years of lost work time.

## Wastewater Treatment Plant

Hazardous material dumped down the drain will end up in the on-site septic system or wastewater treatment plant depending on which system is employed. HHW can enter into wastewater treatment systems through its intended use or as a disposal method. Local governments usually prohibit disposal of HHW into stormwater drains. Recommended disposal may depend on the product and the industry. Some may be dumped down the drain with lots of running water, while others should be kept for collection.

Conventional wastewater treatment plants combined physical, chemical, and biological treatment methods depending on the nature of the pollutants and desired level of removal. Modern wastewater treatment process consists of four levels, including preliminary, primary, secondary, tertiary, or advanced treatment, in addition to the solid waste management. Preliminary and primary treatments are mainly physical/mechanical (screening and gravity settling), while secondary and tertiary treatments use combination of biological, physical, and chemical treatment process. Preliminary treatment removes larger inorganic materials and floating particles, primary treatment removes a major portion (50–60%) of suspended solids from raw wastewater, and secondary treatment process removes organic matters and suspended solids. Secondary treatment usually consists of biological treatment of wastewater. Most of the WWTPs use aerobic activated sludge process for secondary treatment. The objectives of secondary treatment are to reduce BOD and SS of the effluent to an acceptable level according to the discharge regulation. In some cases, nutrient removal may be also an objective of secondary treatment. Biological treatment processes rely upon the ability of the organisms to utilize the contaminants as substrates and results in the generation of new biomass and biodegradation by-products.

Lye and bleach found in cleaning products and other hazardous components can hinder the bacteria that are utilized in the biological treatment processes and will significantly affect the process efficiency. This will cause wastewater to pass through the system without treatment and ultimately will reach the groundwater and/or surface water. This can contaminate aquatic life; nitrates, and phosphates can cause eutrophication (algal bloom), leading to the use of more herbicides for control.

Excess loading of nutrients like nitrates and phosphates results in the uncontrolled growth of phytoplanktons and macrophytes. The growth and subsequent death of these organisms form a greenish slime layer at the surface of water bodies. This slime layer reduces the amount of sunlight that can penetrate through and the oxygen that can be replenished into the water. In addition, the excessive growth causes high competition for resources among aquatic organisms and death such that the biodiversity in the water body may be severely affected over time. This is the water pollution phenomenon known as eutrophication. Aside from the negative effects on water esthetics, eutrophication can hamper recreation activities, navigation, and aquatic life.

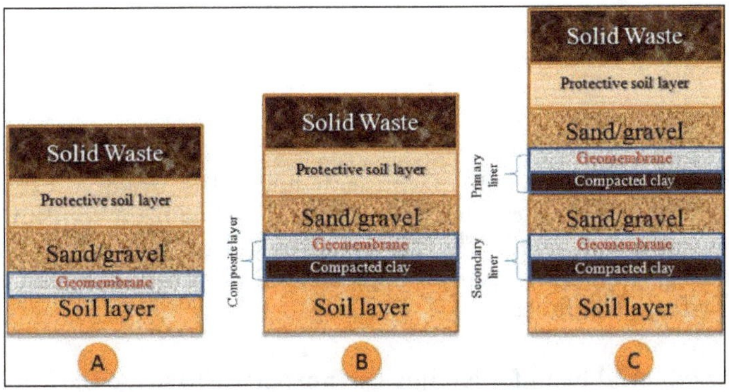


(A) Single liner, (B) composite liner, and (C) double liner system.

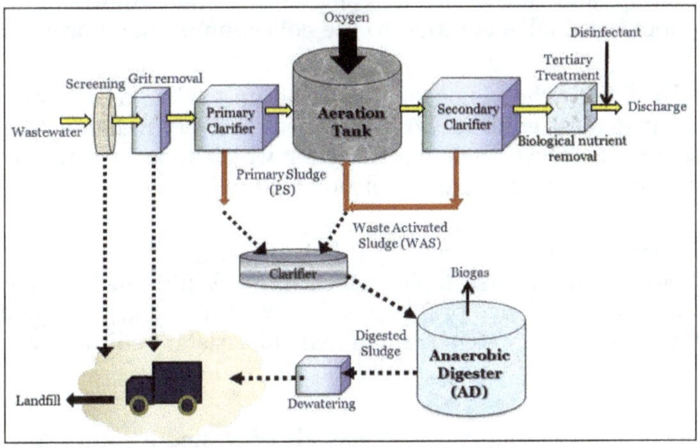


Typical municipal wastewater treatment process.

On the other hand, heavy metals are toxic, persistent, and mobile and tend to accumulate. They generally have very low acceptable concentrations in drinking water standard. In WWTP, low-concentration volatile solvents can evaporate from the aeration tank and become air pollutants. However, high concentrations, acids, bases, poisons, and solvents can affect the WWTP workers' safety and effluent quality and contaminate the sludge. Even if the wastewater flow does not contain HHW, leachate from landfills and combined sewer flow can introduce contaminants from pesticides and motor oil, which originate in the households of which even a small amount of pesticide concentration can cause a WWTP to fail toxicity test.

## Recycling Centers

Majority of the e-waste collected in the US and other developed countries end up in developing countries in Asia and Africa, which often have less than adequate concern for the environmental impacts of the primitive recycling activities that are conducted. Illegal e-waste recycling activities in Guiyu, China, have led to the release of hazardous chemicals into the environment. Harmful concentrations of heavy metals and compounds such as polybrominated diphenyl ethers (PBDEs) were reported in local children and workers of the recycling facilities likely due to open dumping activities that contaminated the soil and river sediments. Polychlorinated biphenyls (PCBs) released during manual dismantling of electronics and from open combustion of the waste material resulted in the presence of significant concentrations in the local residents as a result of bioaccumulation in fish and inhalation.

# Permissions

All chapters in this book are published with permission under the Creative Commons Attribution Share Alike License or equivalent. Every chapter published in this book has been scrutinized by our experts. Their significance has been extensively debated. The topics covered herein carry significant information for a comprehensive understanding. They may even be implemented as practical applications or may be referred to as a beginning point for further studies.

We would like to thank the editorial team for lending their expertise to make the book truly unique. They have played a crucial role in the development of this book. Without their invaluable contributions this book wouldn't have been possible. They have made vital efforts to compile up to date information on the varied aspects of this subject to make this book a valuable addition to the collection of many professionals and students.

This book was conceptualized with the vision of imparting up-to-date and integrated information in this field. To ensure the same, a matchless editorial board was set up. Every individual on the board went through rigorous rounds of assessment to prove their worth. After which they invested a large part of their time researching and compiling the most relevant data for our readers.

The editorial board has been involved in producing this book since its inception. They have spent rigorous hours researching and exploring the diverse topics which have resulted in the successful publishing of this book. They have passed on their knowledge of decades through this book. To expedite this challenging task, the publisher supported the team at every step. A small team of assistant editors was also appointed to further simplify the editing procedure and attain best results for the readers.

Apart from the editorial board, the designing team has also invested a significant amount of their time in understanding the subject and creating the most relevant covers. They scrutinized every image to scout for the most suitable representation of the subject and create an appropriate cover for the book.

The publishing team has been an ardent support to the editorial, designing and production team. Their endless efforts to recruit the best for this project, has resulted in the accomplishment of this book. They are a veteran in the field of academics and their pool of knowledge is as vast as their experience in printing. Their expertise and guidance has proved useful at every step. Their uncompromising quality standards have made this book an exceptional effort. Their encouragement from time to time has been an inspiration for everyone.

The publisher and the editorial board hope that this book will prove to be a valuable piece of knowledge for students, practitioners and scholars across the globe.

# Index


**A**

Adsorption Capacity, 62

Alkalinity, 42, 67

Anaerobic System, 100

**B**

Biocidal Action, 36

Biodegradable Waste, 12-14, 129, 185

Biofilms, 101

Biological Treatment, 61, 98-101, 132, 204

Biomedical Waste, 2

Bioremediation, 61, 103-104

Bisphenol A, 15, 37

Brackish Water, 63, 66

**C**

Cadmium, 2, 10, 18, 20, 23-25, 42, 68-69, 71-72, 74, 116, 192-193

Carbon Adsorption, 61

Carbon Monoxide, 16, 39, 68, 71, 77, 80, 203

Characteristic Wastes, 3

Compressive Strength, 105-107

Corrosive Wastes, 4

**D**

Dimethyl Arsenic Acid, 22

Dioxins, 14, 16, 67-68, 71-72, 80, 82, 111, 126, 193

Distillation, 8-9, 57, 61, 64, 96, 127

**E**

Electro Dialysis, 61

Enthalpy, 88, 90-91, 93-94

**F**

Flocculation, 57, 64-65

Fluorescent Tubes, 10, 117

Fly Ash, 70, 73-74, 104, 106, 126, 171, 204

Fuel Fabrication, 54, 57-58, 120

**G**

Gasification, 68-82, 85-95, 108, 133, 141

Genotoxic Waste, 2

Glutathione Peroxidase, 29, 32

Greenhouse Gas, 14-15, 72-73, 149

**H**

Hazardous Waste Incineration, 95

Hexavalent Chromium, 25

**I**

In Situ Biodegradation, 100

Industrial Wastes, 1, 4-5

**K**

K-listed Wastes, 9

**L**

Leachate, 5, 46, 48, 75, 79, 111, 114, 122-123, 202-203, 205

Lipid Solubility, 19, 31

**M**

Medical Waste, 48-49, 79, 81-82, 85-86

Mercury, 2, 10-11, 18, 20, 30-33, 42, 49, 68-69, 71, 74, 103, 112, 114, 116-117, 122, 192-193, 200, 203

Methane, 14, 68, 73, 75, 79-80, 160

Methylene Chloride, 6, 8, 140, 195, 203

Microbial Sludge, 101

Mixed Hazardous Waste, 12

Molybdenum, 19

Municipal Solid Waste, 13, 42, 49, 68-69, 72, 74, 77, 110-111, 162

**N**

Nickel, 19-20, 25, 71-72, 116, 167

Nitrogen Oxide, 16

Non-biodegradable Waste, 12-14, 129

Noncombustible Waste, 57-58

Nuclear Fission, 53-54

**O**

Ocean Dumping, 54

Oxidative Stress, 19-20, 22, 25, 30-32, 153

**P**

Physicochemical Treatment, 96

Plasma Gasification, 77-81

Plasma Torch, 77, 81, 83, 85, 88, 93

Plastic Pollution, 16-17, 38-39

Plutonium, 51, 55-59, 119-121

**R**
Radioactive Effluents, 53
Radioactive Waste, 2-3, 41, 50-55, 57-59, 119
Reactive Oxygen Species, 20, 29, 31-32
Resin Adsorption, 61-62
Reverse Osmosis, 61, 63, 66

**S**
Saline Water, 66
Sanitary Sewer, 109-110
Sedimentation, 61-66, 87, 96, 99, 101
Sewage Sludge, 78, 82, 86-95, 116
Soil Erosion, 19-20
Soil Pollution, 13, 127, 141
Solid Waste Landfill, 74
Solvent Extraction, 56, 58, 61, 64, 66, 119-120
Solvent Mixture, 6-7
Stoichiometric Mode, 89-92
Styrene Monomer, 34-36, 40, 139, 147
Sulfhydryl Reactivity, 31
Sulfur Dioxide, 68, 71

**T**
Toluene, 7-8, 10, 139, 144, 147, 152, 195
Trickling, 101
Tropospheric Ozone, 39

**U**
Uranium, 20, 51, 54-55, 57, 119-121, 200

**V**
Vinyl Chloride, 15, 43, 203
Vitrification, 87, 89, 91, 93, 95, 104
Volatile Elements, 3
Volatile Organic Compounds, 68, 104-105, 107, 203

**W**
Water Vapor, 82, 91, 94

www.ingramcontent.com/pod-product-compliance
Lightning Source LLC
LaVergne TN
LVHW080650210926
848485LV00003B/356